Key:

□ Caldera
● Subaerial Volcano
◉ Submarine Volcano

VOLCANIC LANDFORMS

and

SURFACE FEATURES

VOLCANIC LANDFORMS AND SURFACE FEATURES

A Photographic Atlas and Glossary

Edited by

Jack Green

McDonnell Douglas Corporation
Douglas Advanced Research Laboratories
Huntington Beach, California

and

Nicholas M. Short

Earth Observations Branch
National Aeronautics and
Space Administration
Goddard Space Flight Center
Greenbelt, Maryland

SPRINGER-VERLAG NEW YORK · HEIDELBERG · BERLIN
1971

© 1971 by Springer-Verlag New York Inc.
Library of Congress Card Number 70-144791
Printed in Japan

ISBN: 3-540-05328 X Springer-Verlag Berlin Heidelberg New York
ISBN: 0-387-05328 X Springer-Verlag New York Heidelberg Berlin

CONTENTS

INTRODUCTION

This book, conceived by N.M.S., is patterned after *The Atlas and Glossary of Primary Sedimentary Structures* by F. J. Pettijohn and P. E. Potter (Springer-Verlag New York, Inc.). We introduce this atlas with a chapter by the late Arie Poldervaart treating the principal concepts of volcanoes as landforms, followed by a main section of photographs of volcanic structures and features arranged in 198 Plates, and then conclude with an updated glossary of terms associated with volcanology, its processes and products.

The atlas is, in a sense, an outgrowth of the expanding interest in volcanology recently stimulated by the exploration of neighboring planetary bodies in the solar system. Ranger, Surveyor, Orbiter, and Apollo photographs have provided strong evidence for the existence of volcanic activity on the Moon. Even the 1965 and 1969 Mariner television pictures sent from Mars suggest that volcanism is not confined to the Earth and its satellite but may be a generally operative planetary process. In anticipation of the ever-increasing need to have improved visual models of volcanological features as subsequent landings on the lunar surface take place, photographs of many volcanic landforms and textures have been compiled so as to better interpret high and low altitude and ground photographs of the inner planets, including our Earth as it is resurveyed from space. Because of this emphasis, we are including a few representative photos of this type.

However, we recognize that the vast bulk of photographs, and subjects depicted, of volcanological character obtained during this century are strictly earthbound in source, scope, and application. Thus, the photos that we selected to illustrate volcanic features are, of necessity, taken from regions of our own planet, some of which have become classic localities for display of the principal aspects of volcanism. This phase is related to the second prime objective for preparing this atlas, namely to assemble into a single source book a photographic record of nearly all volcanic surface features described during the development of volcanology so that future workers on terrestrial problems can refer to these photos for comparative or illustrative purposes.

Also, we hope that this atlas will serve as an aid to those engaged in learning or teaching the fundamentals of geology and its subfields, such as petrology or geophysics. To this end we have attempted to create a book simple and general enough to be useful even at the secondary school level, but with sufficient detail and rigor to be acceptable to both students and professors in the universities. Further, we expect this book to be useful to geomorphologists concerned with landforms associated with volcanic terrains. Finally, by keeping this atlas as non–technical as possible, adding the Poldervaart chapter which reviews the general principles of volcanology from the viewpoint of its influence on the Earth's scenery, and including a comprehensive glossary, we trust that this book will have broad appeal to the many non–professionals who are interested in volcanic phenomena only as a natural wonder.

The atlas is designed, therefore, to reach out to a wide audience. Consider a panoramic photograph of the Mayon volcano in the Philippines or Fujiyama in Japan. The elementary school child sees the simple geometry of the volcano; the high school student will note the symmetry of the volcano and the flank flow patterns; and the college student observes its slope angle as related to host rock composition, or the morphological details of the summit crater, or an adventive cone. A graduate student specializing in volcanology may perceive the relationship of the volcano to the regional tectonics. He would observe fissure-aligned solfataras, the spacing of flow corda, and the terminal angle of the flow fronts. He might wonder why isolated stratovolcanoes seemingly do not occur on the Moon. In the field, the working scientist would combine vertical and oblique photography at different scales and wavelengths with analytical equipment. He would hope to measure the quantity and composition of gases or effusive solids released in recent eruptions. If possible, he would determine surface tilting of volcano slopes as a function of heat flow. He would station seismographs to record variations in shallow depth shocks as possible precursors of an impending eruption. The scientist would speculate on the energy required to form these volcanoes or maintain their activity and would attempt to synthesize all available data into a comprehensive hypothesis for the physico-chemical behavior of these natural outlets of magmatic materials from the deeper crust or mantle.

The arrangement of the plates is from large to small features beginning with volcano-tectonic depressions to calderas to volcanoes. Then, smaller features such as maars and domes are documented. Following this are photographs of flow structures, hornitos, fumaroles, volcanic bombs, and the like. At a still finer scale, various surface features are presented such as lava coils, pipe vesicles, and finally characteristic field textures of volcanic rocks are exemplified. In the concluding section, we have added eight Lunar Orbiter photos selected to show established or probable volcanic features visible on the Moon's surface.

Wherever possible, we have placed two photos to a page, with the captions and photo credits for these appearing on the opposite page. In some instances, we have put as many as four photos to a page or, in a few cases, we have included a photo as an inset within a larger photo. In making the choice in favor of more multiple photo plates, we were guided by a single principle, to illustrate as many different aspects of volcanology as practical. Sometimes, the same feature or phenomenon is depicted by several photos, but we felt in these instances that the apparently repetitive coverage was justified by the variations which characterize the subjects being described.

As far as possible, the captions contain not only identification and location of features shown in a photo along with appropriate facts or discussion on particular aspects of these features, but also additional statements about the nature, origin, processes, or characteristics of the features as general classes or categories.

The reader unfamiliar with terms or concepts introduced in a caption will find almost all of them defined in the Poldervaart chapter or the Glossary.

Problems of nomenclature are often troublesome in volcanology. For example, what sets of characteristics distinguish the summit pit of a volcano from a caldera? Or, to provoke controversy, how does the summit pit of a large volcano evolve into a caldera? A glossary, of course, cannot provide fully satisfactory answers to such questions, but it does serve to summarize the current thinking of the experts on the general, and sometimes precise, meanings applied to terms relating to forms and features, processes and rock products of volcanoes.

The glossary developed for this Atlas has been abstracted from the revision now being made for the third edition of the *Glossary of Geology and Related Sciences* published by the American Geological Institute. We are indebted to that organization, and especially to the Glossary Editor, Dr. Robert McAfee, for allowing us to have access to the file cards for terms in volcanology and pyroclastic rocks and for permission to use selected definitions from these cards in compiling our glossary.

This glossary has been arranged alphabetically with terms for volcanic landforms and processes interspersed with those for the pyroclastic rocks. We have omitted most of the terms more properly associated with intrusive igneous rocks with the exception of a few terms such as "dike phase" or "magma chamber" which have an obvious connection with extrusive rocks.

For many entries, more than one definition is cited, with the reference for each keyed to a source list at the end of the glossary. We have omitted multilanguage definitions, but have included some commonly used terms derived from languages other than English. The reader interested in the French, German, Spanish, and Dutch equivalents to most of the terms presented in our glossary should consult A.A.G. Schieferdecker's book on *Geological Nomenclature* (Royal Geological and Mining Society of the Netherlands, J. Noorduijn en Zoon N. V., 1959).

Despite the care exercised in selecting the specific definitions incorporated in the glossary, many of these are still in need of further refinement. For space exploration it is even more pertinent to have these definitions developed and keyed to representative photographs of terrestrial examples in order to better describe forms and textures of volcanic structures and rocks which may exist on Mercury, Venus, the Moon, and Mars. We hope that this attempt to present a fairly complete volcanological glossary will, in effect, also point out the inconsistencies, ambiguities, and redundancies of terminology that need to be resolved.

Mention should here be made about the variable quality of the photographs used in this book. Non-professional organizations who spend large sums of money on photography, not being aware of the significance of features of volcanological interest, often photograph these objects under conditions adverse to the definition of scientific details. Among geoscientists, factors contributing to less-than-professional photography is the lack of time and/or the expense of waiting for the right weather or best sun angle. Often it is impossible to photograph transient volcanological phenomena at the right time or from the "best" view site. Furthermore, many geoscientists use color film which loses much definition on conversion to black and white. Also to add "venerability" to the book we have intentionally included some classic photographs even if these were not photographically superior.

We wish to express our sincere gratitude to NASA's Goddard Space Flight Center, Greenbelt, Maryland and to the Douglas Advanced Research Laboratories (McDonnell Douglas Corp.) of Huntington Beach, California. Each of these organizations contributed in various ways to the preparation of this atlas. We would also like to thank Drs. R. L. Smith, R. Bailey, and R. S. Fiske of the U.S. Geological Survey, and Drs. T. Simkin and W. Melson of the Smithsonian Institution in Washington, D.C. for their many suggestions concerning the style and content of the atlas as it evolved.

Finally, in compiling a group of photographs of the scope attempted, we have had the pleasure of contacting many other active workers in the field of volcanology. To these people we extend our gratitude. Organizations not specifically oriented toward volcanology also opened their photo catalogs to us. In particular, we are indebted to the Geophysical Laboratory of Carnegie Institution for permitting us to publish some classics. Each contributor is cited in the caption of a specific photograph. However, in the following list we doubly emphasize that whatever excellence this book may have rests on those listed in the Contributors section.

October 1970   J. Green
                      N. M. Short

LIST OF PLATES

Section C: VOLCANOES

Section D: INTERNAL STRUCTURE OF VOLCANOES

Section E: CRATERS AND MAARS

Section F: TUFF AND CINDER CONES

Section G: DOMES AND LACCOLITHS

Section H: SPINES, NECKS AND DIATREMES

Section I: DIKES AND SILLS

Section J: ALIGNMENTS OF VOLCANIC FEATURES

Section K: GENERAL CHARACTERISTICS OF FLOWS

Section L: MUDFLOWS AND LAHARS

Section M: FEATURES OF PYROCLASTIC DEPOSITS

Section N: STRUCTURES ON LAVA FLOW SURFACES

Section O: UPSWELLINGS AND MARKINGS ON LAVA SURFACES

Section U: EROSION FEATURES IN VOLCANIC ROCKS

LIST OF CONTRIBUTORS

Dr. S. Aramaki
Earthquake Research Institute
University of Tokyo
Tokyo, Japan

Dr. Roy A. Bailey
U.S. Geological Survey
Washington, D.C.

Dr. P. E. Baker
Dept of Geology and Mineralogy
University of Oxford
Parks Road, Oxford, England

Mr. S. N. Beatus
New Zealand Geological Survey
Lower Hutt, New Zealand

Dr. R. G. Bowen
Dept. of Geology and Mineral Industries
State of Oregon
Portland, Oregon

Dr. G. M. Brown
Department of Geology
University of Durham
South Road, Durham City, England

Dr. Fred M. Bullard
Department of Geological Sciences
The University of Texas
Austin, Texas

Mrs. W. S. Cameron
Division of Theoretical Studies
NASA, Goddard Space Flight Center
Greenbelt, Maryland

Dr. A. H. Chidester
Branch of Astrogeologic Studies
U.S. Geological Survey
Flagstaff, Arizona

Dr. U. Clanton
National Aeronautics and Space Administration
Manned Spacecraft Center
Houston, Texas

Dr. Howard A. Coombs
Department of Geology
University of Washington
Seattle, Washington

Dr. A. Corcoran
Dept. of Geology and Mineral Industries
State of Oregon,
Portland, Oregon

Dr. R. W. Decker
Department of Earth Sciences
Dartmouth College
Hanover, New Hampshire

Prof. M. Derruau
Dept. de Géographie
Université de Clermont-Ferrand

Dr. R. R. Dibble
Department of Geology
Victoria University of Wellington
Wellington, New Zealand

Dr. Gordon Eaton
Regional Geophysics Branch
U.S. Geological Survey
Denver, Colorado

Dr. Wolfgang E. Elston
Department of Geology
University of New Mexico
Albuquerque, New Mexico

Embassy of Ecuador
Washington, D.C.

Embassy of France and French Cultural
Services, Washington, D.C.

Dr. R. V. Fisher
Department of Geology
University of California
Santa Barbara, California

Dr. R. S. Fiske
U.S. Geological Survey
Washington, D.C.

Dr. Theodore Foss
Geology and Geochemistry Branch
NASA, Manned Spacecraft Center
Houston, Texas

Dr. Helen Foster
U.S. Geological Survey
Menlo Park, California

Dr. Jules Friedman
U.S. Geological Survey
Washington, D.C.

Mrs. Kerstin K. Friedman
Fairfax Virginia

Geophysical Laboratory of Carnegie
Institution of Washington
Washington, D.C.

Dr. Ian L. Gibson
Department of Earth Sciences
The University,
Leeds, England

Dr. G. S. Gorshkov
Institute of Volcanology
Academy of Sciences of the USSR
Moscow, U.S.S.R.

Dr. Ronald Greeley
Planetology Branch
NASA, Ames Research Center
Moffett Field, California

Dr. Warren Hamilton
U.S. Geological Survey
Federal Center
Denver, Colorado

Hansa Luftbild, Gmbh
Münster, W. Germany

Hawaii Visitors Bureau
Honolulu, Hawaii

Dr. Carter Hearn
U.S. Geological Survey
Washington, D.C.

Mr. D. L. Homer
New Zealand Geological Survey
Lower Hutt, New Zealand

Dr. Keith Howard
U.S. Geological Survey
Menlo Park, California

Mr. J. Hughes
Office of Naval Research
Washington, D.C.

Hunting Overseas Surveys
London, England

Indonesian Department of Information,
Jakarta, Indonesia

Institute of Geological Sciences
Geological Survey and Museum
Exhibition Road, South Kensington
London, England

Japan Air Lines
Washington, D.C.

Mr. Sigurgeir Jonasson
Box 25
Vestmannaeyjar, Iceland

Dr. F. Jones
U.S. Geological Survey
Washington, D.C.

Mr. R. Kachadoorian
Alaskan Geology Branch
U.S. Geological Survey
Menlo Park, California

Dr. Gerard P. Kuiper
Lunar and Planetary Laboratory
The University of Arizona
Tucson, Arizona

Dr. T. J. W. Lee, III
U.S. Geological Survey
Washington, D.C.

Mr. P. W. Lipman
U.S. Geological Survey
Denver, Colorado

Dr. Paul D. Lowman
Planetology Branch
NASA, Goddard Space Flight Center
Greenbelt, Maryland

Dr. G. A. Macdonald
Department of Geosciences
University of Hawaii
Honolulu, Hawaii

Dr. Warren Manspeizer
Geology Department
Rutgers. The State University
Newark, New Jersey

Dr. G. J. H. McCall
University of Western Australia
Nedlands, Australia

Dr. J. S. McCauley
Branch of Astrogeologic Studies
U.S. Geological Survey
Menlo Park, California

Dr. Thomas R. McGetchin
Dept. of Earth Sciences
Mass. Inst. of Technology
Cambridge, Mass.

Dr. William E. Melson
Division of Petrology and Volcanology
U.S. National Museum
Washington, D.C.

Dr. James G. Moore
U.S. Geological Survey
Menlo Park, California

National Aeronautics and Space
Administration
Langley Field and
Goddard Space Flight Center

National Park Service
Branch of Still Pictures
Washington, D.C.

Mr. Tad Nichols
5545 Camino Escuela
Tucson, Arizona

Dr. R. L. Nichols
Eastern Kentucky University
Richmond, Kentucky

Dr. Yoshio Oba
Dept. of Geology and Mineralogy
Hokkaido University
Sapporo, Japan

Mr. F. O'Leary
New Zealand Geological Survey
Lower Hutt, New Zealand

Dr. Willard H. Parsons
Department of Geology
Wayne State University
Detroit, Michigan

Mr. A. Patnesky
National Aeronautics and Space Administration
Manned Spacecraft Center
Houston, Texas

Dr. Dallas L. Peck
U.S. Geological Survey
Washington, D.C.

Dr. Angelo Pesce
P. O. Box 661
Tripoli, Libya

Brig. General Giorgio Pocek
Office of the Air Attache
Italian Embassy
Washington, D.C.

Dr. Howard Powers
Hawaiian Volcano Observatory
Hawaii

Dr. Martin Prinz
Dept. of Geology
University of New Mexico
Albuquerque, N. Mexico

Mr. D. Reeser
Hawaiian Volcano Observatory
Hawaii

Dr. J. S. Rinehart
Institutes for Environmental Research
Environmental Science Services Admin.
Boulder, Colorado

Dr. D. J. Roddy
Branch of Astrogeologic Studies
U.S. Geological Survey
Flagstaff, Arizona

Royal Air Force, England

Dr. C. J. Schuberth
American Museum of Natural History
New York, N.Y.

Dr. John S. Shelton
1100 Oxford Avenue
Claremont, California

Dr. Thomas Simkin
Division of Petrology and Volcanology
U.S. National Museum
Washington, D.C.

Dr. Robert L. Smith
U.S. Geological Survey
Washington, D.C.

South Dakota Department of Highways
Vermillion, South Dakota

Dr. H. G. Stephens
U.S. Geological Survey
Washington, D.C.

Dr. R. Sutton
U.S. Geological Survey
Flagstaff, Arizona

Mr. H. Takeshita
Moji High School
Moji-ku, Kitakyushu-shi, Japan

Dr. Bruce N. Thompson
New Zealand Geological Survey
Otara, New Zealand

Dr. Sigurdur Thorarinsson
Museum of Natural History
Reykjavik, Iceland

U.S. (Army) Air Force
Washington, D.C.

U.S. Geological Survey
Map Information Office
Washington, D.C.

U.S. Information Office
Washington, D.C.

U.S. Navy
Washington, D.C.

Dr. Roland von Huene
U.S. Geological Survey
Menlo Park, California

Dr. Donald E. White
U.S. Geological Survey
Menlo Park, California

Dr. Howard Wilshire
U.S. Geological Survey
Menlo Park, California

Dr. Kenzo Yagi
Dept. of Geology and Mineralogy
Hokkaido University
Sapporo, Japan

Dr. I. Yokoyama
Geophysical Institute
Hokkaido University
Sapporo, Japan

VOLCANIC LANDFORMS
and
SURFACE FEATURES

VOLCANICITY AND FORMS OF EXTRUSIVE BODIES[1]

Arie Poldervaart

1. *Types of Eruptions*: Extrusive or effusive igneous rocks have consolidated from magma poured out or blown out over the earth's surface. Two major types of volcanic activity are recognized: fissure eruptions and central eruptions. The broader classification of volcanoes according to characteristics of their eruptions, proposed by Lacroix in 1908 and formalized by Sapper in 1931 (2), is summarized in Table I. The general worldwide distribution of volcanic belts, with specific location of well-known individual volcanoes, is depicted in Figure 1. (see endpapers)

2. *Fissure Eruptions*: Eruptions of this type are concentrated along long, narrow fissures. Lava may issue from the fissures and spread to form lava plateaus (Plates 83, 84), or the eruptions may be mainly explosive and result in ash flows that form plateaus of fragmental volcanic rocks (3). Many fissure eruptions are associated with basaltic lava flows, hence the names *flood basalts* or *plateau basalts* (Plate 97) (4). Basaltic melts are rather fluid and crystallize to rocks containing mainly calcic plagioclase and pyroxene, with or without olivine. Because of the low viscosity of basaltic magma, individual flows average 50 feet* or less in thickness, but may extend over large areas.

Flows several hundred feet thick are rare and result from ponding of lava in depressions. Small volcanic cones are commonly aligned along the fissures (Plates 84B, 87). Large basaltic fissure eruptions have not been prominent in historical times. The best known is the 1783 eruption along the 20-mile long Laki fissure in Iceland, from which some three cubic miles of basalt poured over the surface (Plate 85A) (5). Traces of another recent eruption are preserved along a fissure in the Craters of the Moon National Monument (6) (Plate 83A) in Idaho, which yielded the last lava contribution to the vast Snake River plateau. In the geologic past, basaltic fissure eruptions were more important and formed immense lava plateaus, thousands of feet thick and thousands of square miles in area. Examples of this are the late Tertiary to Pleistocene Snake River and Columbia River Plateaus (combined area nearly 200,000 square miles) of the northwestern United States (Plate 96, 97) and the late Cretaceous to Eocene Deccan basalts of peninsular India which have a similar areal extent. Fissure eruptions of ash flows are associated with magmas that are more silicic than basalt. Examples are discussed in Sections 6 and 8.

3. *Central Eruptions*: Volcanic action of this type is concentrated at a central vent and gives rise to volcanic cones and domes, with or without summit craters. Volcanoes may be *single* (Plates 28–32), form *clusters* (Plates 52, 56), or be distributed in *chains* (Plates 33, 85). Mount Kilimanjaro on the Tanganyika-Kenya border is an example of an isolated, single, giant volcano. Volcanic clusters are found near Mexico City and Naples, Italy. The volcanic chains of the Japanese and Indonesian

---

[1]The numbers in parentheses are keyed to individual or groups of references which are listed at the end of this article.

*Although the metric system has been used exclusively throughout the remainder of this atlas, the English system of numbers and units originally employed by Poldervaart in this article has been retained here in order to avoid redrafting of those illustrations that have scales, etc. in the English system.

Archipelagos are also well known. The famed circum-Pacific "Belt of Fire" is an example of a huge volcanic belts, consisting of numerous volcanic chains. Volcanic chains are associated with belts of intense seismic disturbances, and also commonly mark belts of strong gravity anomalies. Volcanic cones are subdivided into lava cones, pyroclastic cones, and composite cones.

Lava cones ((Plate 40) are built mainly of lava flows that issued from the central vent. *Spatter*

TABLE I
TYPES OF VOLCANIC ERUPTIONS*

Type	Characteristics
1. Icelandic	Fissure eruptions, releasing free flowing (fluidal) basaltic magma; quiet, gas-poor; great volumes of lava issued, flowing as sheets over large areas to build up plateaus (Columbia).
2. Hawaiian	Fissure, caldera, and pit crater eruptions; mobile lavas, with some gas; quiet to moderately active eruptions; occasional rapid emission of gas-charged lava produces fire fountains; only minor amounts of ash; builds up lava domes.
3. Strombolian	Stratocones (summit craters); moderate, rhythmic to nearly continuous explosions, resulting from spasmodic gas escape; clots of lava ejected, producing bombs and scoria; periodic more intense activity with outpourings of lava; light-colored clouds (mostly steam) reach upwards only to moderate heights.
4. Vulcanian	Stratocones, (central vents); associated lavas more viscous; lavas crust over in vent between eruptions, allowing gas buildup below surface; eruptions increase in violence over longer periods of quiet until lava crust is broken up, clearing vent, ejecting bombs, pumice and ash; lava flows from top of flank after main explosive eruption; dark, ash-laden clouds, convoluted, cauliflower shaped, rise to moderate heights more or less vertically, depositing ash along flanks of volcano (Note: pseudo-Vulcanian eruption has similar characteristics but results when other types (e.g., Hawaiian) become phreatic and produce large steam clouds, carrying fragmental matter).
5. Vesuvian	More paroxysmal than Strombolian or Vulcanian types; extremely violent expulsion of gas-charged magma from stratocone vent; eruption occurs after long interval of quiescence of mild activity; vent tends to be emptied to considerable depth; lava ejects in explosive spray (glow above vent), with repeated clouds (cauliflower) that reach great heights and deposit ash.
6. Plinian	More violent form of Vesuvian eruption; last major phase is uprush of gas that carries cloud rapidly upward in vertical column for miles; narrow at base but expands outward at upper elevations; cloud generally low in ash.
7. Peléan	Results from high viscosity lavas; delayed explosiveness; conduit of stratovolcano usually blocked by dome or plug; gas (some lava) escapes from lateral (flank) openings or by destruction or uplift of plug; gas, ash, and blocks move downslope in one or more blasts as nuées ardentes or glowing avalanches, producing directed deposits.

*Abstracted and modified from Chapter XII (pp. 305–310) of *Principles of Physical Geology* by A. Holmes, 2nd ed., Ronald Press, with additional data from *Volcanoes: In History, In Theory, In Eruption*, by F. M. Bullard, Univ. of Texas Press.

cones (Plates 87, 123–124) are small, steep-sided cones (less than 100 feet high), formed by spatter from ebullient lava. *Hornitos* or *driblet cones* (Plates 125–127) are also small and steep-sided but are formed by squeezing up of viscous lava. Larger lava cones consist exclusively of basaltic flows and are of two main types: Icelandic and Hawaiian. Both are characterized by gentle slopes of 6°–12°, or about 1,000 feet per mile. Few *Icelandic cones* are more than 3,000 feet high and most have a small summit crater, commonly with a raised rim exemplified by Skjaldbreid (Plate 30B). *Hawaiian cones* are very large, more than 5,000 feet in height and scores of miles across at their base (7). They are dome- and shield-shaped volcanic complexes, rather than cone-shaped structures, with slopes diminishing to 300 feet per mile or less near the summit. The type example is Mauna Loa, Hawaii (Plate 26A). which rises approximately 30,000 feet above the ocean floor (13,680 feet above sea level), and has a nearly flat summit containing the large central crater, Mokuaweoweo (Plate 26B). The volume of this giant volcanic complex is estimated at 10,000 cubic miles, yet individual flows average only 10 feet thick (8). Shield volcanic complexes are especially characteristic of the central Pacific Ocean.

Pyroclastic cones (*ash cones or tuff cones*) (Plates 50–52) consist predominantly of fragmental materials produced by explosive volcanic activity (9). Slopes of the cones are largely determined by the angle of repose of the ejected material. Fine volcanic ashes are scattered farthest from the central vent, forming slopes of about 10° near the base of the cone. Coarser fragments build the higher slopes at angles of 20°–30°, while even larger fragments near the summit support slopes exceeding 30°. Such variations in the angle of repose of ejecta partly explain the classic concave volcano profile; a contributing factor is the dishcarge of lava from fissures near the base of the cone. Large composite volcanoes may start as pyroclastic cones; Parícutin in Mexico (Plates 1, 179A), now a dormant volcano nearly 1,300 feet high, built a 500-foot pyroclastic cone during its first month in 1943 (10). Pure pyroclastic cones are relatively small, generally less than 2,000 feet high. *Cinder cones* (*scoria mounds*) (Plates 54–58) are small, steep-sided pyroclastic cones built of fragments or volcanic clinkers.

Composite cones (*stratovolcanoes*) (Plates 28–36) are stratified and consist of both lava flows and beds of pyroclastic materials. Nearly all large volcanic cones on continents are composite but the propor-

tion of lava flows to fragmental materials varies from one volcano to the next. Mount Etna, Sicily (Plate 36), has a high proportion of lava flows and its shape approaches that of a Hawaiian shield volcanic complex. The proportion of lava flows to pyroclastic materials in large Indonesian stratovolcanoes of the Sunda belt is estimated at 1:350.

4. *Other Features of Volcanoes*: The simple cone or dome shape of a large volcano is commonly altered by subsidiary cones on its flanks; these are *parasitic* or *adventive cones* (Fig. 2) (Plate 52B). Faults, fis-

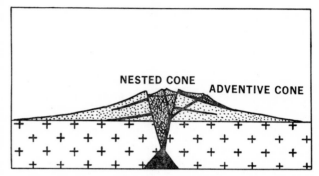

Fig. 2: Composite Volcano with Nested and Adventive Cones (Schematic)

sures, and rifts may also form on the flanks of large volcanoes, and lava can issue from these cracks with all the characteristics of fissure eruptions. The summit crater of a volcano may be *breached* (Plate 62A), portion of the crater walls having been removed by outflow of lava or by explosive activity. A large crater may include one or more *nested cones* or craters (Fig. 2) (Plates 22B, 34A). The remnant of an old crater wall is called a *somma*, after Mount Somma, the ancient rim of Vesuvius which partly encircles the present active cone (11) (Plate 34A).

Viscous lava may flow for short distances and form tongues of small to moderate dimensions. Such small flows or *coulees* are relatively thick and have steep sides, giving distinctly convex cross sections (Plate 93B). They congeal on the steep upper slopes of composite volcanoes (Plate 28); large blocks and red hot talus may break away from the coulees, yielding accumulations of volcanic breccia at lower levels. Highly viscous lava scarcely flows but is mostly squeezed up to form a *cumulo dome* or *tholoid* (12) (Plate 62). Two types of cumulo domes are recognized: *endogenous* domes that grow by expansion from within (Plate 60) and *exogenous* domes which are built by surface effusion, normally from a central summit crater (Plate 63A). Most domes are endogenous in the initial stages of their growth. Through continued internal expansion, the crust becomes intensely fractured and during later stages of growth lava issues from the fractures, the domes then becoming compound endogenous and exogenous in character. Exogenous domes generally have a smooth crust; Grand Sarcoui in the Auvergne district, France, is an example. Puy de Dome also in Auvergne, is an endogenous dome. Many other puys in Auvergne are plug domes or volcanic necks (Plate 70), discussed below. The dome of Mont Pelée, Martinique, is also an endogenous dome. The rate of growth of volcanic domes is exceedingly rapid in comparison to that of stratovolocanoes. The dome of Mont Pelée reached a height of some of 1,200 feet and a diameter of more than half a mile in 1-½ years, despite the fact that it suffered repeatedly from disastrous explosions (13). In one day it increased its height by 75 feet.

Volcanic spines may protrude from the crust or carapace of cumulo domes; the most celebrated example being the 1,000-foot high spine that developed in 1903 on the tholoid of Mont Pelée (Plate 69A). Spines are short-lived features and are destroyed by volcanic explosions. Viscous lava may expand upon extrusion, forming a cumulo dome of spreading form, or the lava may retain its cylindrical shape, producing a variant of cumulo domes, the *plug dome* or *piton* (14) (Fig. 3) (Plates 61A, 69B). Many cumulo domes and pitons are found at Lassen Volcanic National Park (Plate 61B) and among the Mono Craters of California (15) (Plates 65–66). The domes of Lassen Peak and vicinity are among the largest in the world. Lassen Peak appears to have risen about 2,500 feet above the original crater and has a basal diameter of some 1-½ miles.

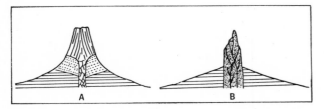

Fig. 3: Types of Lava Domes [Modified from Tanakadate, 1930, p. 704 (14)] A. Cumulo Dome of Spreading Form; B. Plug Dome or Piton

Denudation of a volcanic cone commonly leaves resistant fillings of the former vent standing out in bold relief. (Plate 70). Such cylindrical monoliths are called volcanic necks and are of two main types (16) (Fig. 4) The first type (A) consists of remnants

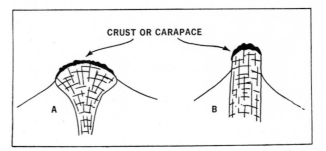

Fig. 4: Types of Volcanic Necks [Modified from Williams, 1936, Fig 4, p. 122 (16)] A. Hopi Neck—Diverging Lava Columns Resting on Inward-dipping Tuffs; B. Navajo Neck—Monolith of Volcanic Breccia Riddled with Dikes

of cumulo domes of spreading form which are of larger diameter than the deeper root portions; this *Hopi type* of volcanic neck has diverging columns of lava resting on inward-dipping tuffs, (Plate 71). The second type (B) consist of remnants of the pipes themselves, the *Navajo type* of volcanic neck which consists of consolidated pyroclastic materials riddled with dikes (Plate 73B). Some volcanic necks form the center of radiating dikes that may stand out in relief and form narrow walls (Plate 72). Many excellent examples of these residual volcanic features are found in the Navajo-Hopi country of Arizona, New Mexico (Plate 73A).

5. *Calderas*: Calderas are large craters (more than a mile across)*. Three main types of calderas are recognized (17): *collapse, explosion*, and *erosion calderas*. Collapse calderas are further subdivided into Krakatau-type or explosion-collapse calderas (Fig. 5A) and Glencoe-type or cauldron-subsidence calderas (Fig. 5B).

In *Krakatau-type calderas*, collapse follows eruption of large amounts of pumice that mantle the flanks of the caldera and the surrounding region. The type example is Krakatau in the Sunda Strait, Indonesia (Plate 4A), which erupted in 1883, ejecting 4½ cubic miles of pumice and subsiding to form a submarine caldera about 4 miles across and 850 feet below sea level at the deepest points (18) (Fig. 6). The active volcanic cones of Danan and Perbuwatan, as well as part of the Rakata cone, disappeared in this catastrophic eruption. In 1927 a new cone, Anak Krakatau (Child of Krakatau) (Plate 4B), emerged above the old eruptive center and this volcano has been intermittently active ever since. Prior to the 1883 eruption Krakatau was already a caldera.

Crater Lake, Oregon (19) (Plate 24) is also a

---

*But see definition by editors in Glossary.

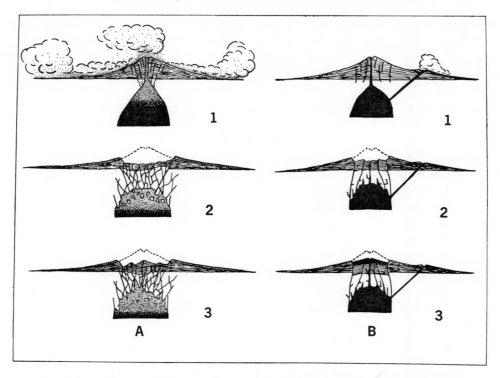

Fig. 5: Types of Collapse Calderas. A. Krakatau Type: (1) Eruptive Phase with Ash Showers and
Nuees Ardentes, (2) Collapse to Caldera, (3) Renewed Activity and Formation of Nested
Cones B. Glencoe Type: (1) Eruptive Phase with Lava Outflow from Adventive Cone,
(2) Collapse to Caldera, (3) Renewed Activity and Formation of Nested Lava Cone.

Krakatau-type caldera, with a diameter of 5½ miles. The original volcanic structure, Mount Mazama, was probably about 12,000 feet high, and erupted and collapsed around 7,300 years ago. In many Krakatau-type calderas, the amount of material ejected is far less than the bulk of material that disappeared. Hence it is believed that much of the space for engulfment was provided by withdrawal of magma at depth. At Mount Mazama, 17 cubic miles of the volcanic complex vanished but only some 7½ cubic miles can be accounted for by the surrounding ejecta.

In a *Glencoe-type caldera* (20) subsidence does not follow pumice eruption, but results from isolation of the central part of the cone by ring fractures and withdrawal of magmatic support by retreat of magma or by lateral intrusion or extrusion. Examples of Glencoe-type calderas are Mokuaweoweo, the summit caldera of Mauna Loa (Plate 26), and Kilauea, also in Hawaii (Plate 27A) (Fig. 7) (21). These calderas have also been called *volcanic sinks*. Smaller equivalents are the Hawaiian *pit craters* (Plates 43, 44), with their steep walls and flat floors of solid or liquid lava. Mokuaweoweo was formed, at least in part, by coalescence of several such pit craters. Another example of a Glencoe-type caldera is the Newberry Volcano, Oregon

(22), which measures more than 4 miles across. In both types of collapse calderas, continued post-caldera activity may result in nested cones and craters, or parasitic rim cones. Hakone Volcano, Japan (23) has an older caldera of the Glencoe-type, within this a younger Krakatau-type caldera, and within the younger caldera a nested, steep-

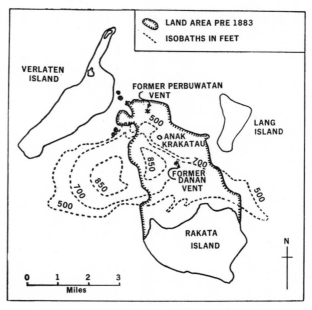

Fig. 6: Krakatoa (Krakatau) Caldera, Sunda Strait, Indonesia.

sided composite volcano and six lava domes. Some calderas are very large; such giant calderas include Valles, New Mexico (24) (Plate 16), 14 miles across, Aira and Kikai, Japan, respectively 14 and 11 miles in diameter, Ngorongoro, Tanganyika, more than 12 miles across, and Lago di Bolsena, Italy and Idjen, Indonesia, both 11 miles in diameter.

Calderas resulting solely from explosive activity are rare, as pure volcanic explosion features are generally less than 1 mile across. Bandai-San, Japan (25) is commonly cited as the type example of an *explosion caldera*. Its eruption in 1888 blew out part of the crater wall, triggering a landslide that destroyed one side of the crater, leaving an amphitheater 8,000 feet across with walls 1,200 feet high, while the debris was scattered over a large mound field at the base. Another example is the 1886 Tarawera eruption in North Island, New Zealand (26) (Plate 86), which blew out the former Lakes Rotomahana and Rotomakiri, leaving an irregularly shaped hollow that was later filled to become the present Lake Rotomahana. In both cases it may be stretching the term "caldera" to include these land forms but true explosion calderas may result from coalescence of two or more explosion craters. In the third type of caldera, the *erosion caldera*, an original crater is enlarged by erosion and retreat of the crater walls. An example is Papenoo Caldera of Tahiti, which measures about 4 miles across.

6. *Other Volcanic Depressions*: Magmas in contact with water-saturated sediments at or below the water table, and gas-charged magmas under shallow cover are highly explosive. Water-related explosive activity is called *phreatic*. Vents drilled through the enclosing rocks by the explosive energy of gas-charged magmas are *diatremes* (27) (Plate 74); the kimberlite pipes of southern and eastern Africa are examples. *Maars* (Plates 46–49) are flat-floored craters of explosion, with diameters much greater than the heights of surrounding rims. Type examples are in the Eifel, Germany. (Plate 46A) Ejected materials may be fragmented country rocks only or of country rocks mixed with pyroclastic materials. Maars, or *embryo volcanoes*, can be dry or water-filled (lakes) and may be without cones but are sometimes surrounded by low tuff rings. Between Lake Edward and Lake George in southwestern Uganda, there are numerous examples of such explosion craters, in two belts north and south of the Kazinga Channel. The maars are surrounded by low rings consisting in part of oolitic calcareous lake sediments and in part by igneous fragments. Generally, these embryo volcanoes are less than one mile across, but coalescence of several explosion craters can produce explosion calderas. The Ubehebe Craters of Death Valley, California (28) (Plate 48B) are another example. Subsidence on the flanks of volcanic cones may produce *sector grabens* or *volcanic rents* (Plate 84B). Large open fissures or *volcanic rifts* also form in volcanic terrain. Examples include the 20 mile-long Eldgja rift of Iceland and

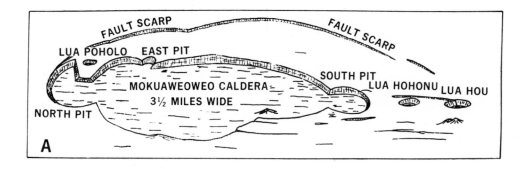

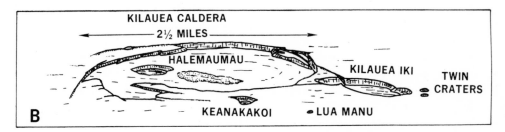

Fig. 7:   Hawaiian Calderas and Pit Craters. A. Mokuaweoweo Caldera; B. Kilauea Caldera.

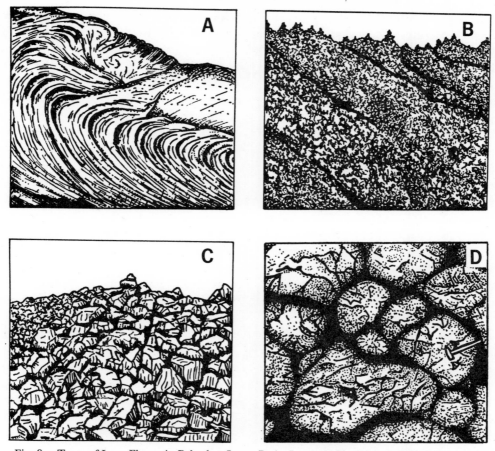

Fig. 8: Types of Lava Flows. A. Pahoehoe Lava; B. Aa Lava; C. Block Lava; D. Pillow Lava.

the Tarawera rift of North Island, New Zealand.

Volcano-tectonic depressions are large, irregular areas of subsidence related to voluminous eruptions. An example is the Lake Toba depression in northern Sumatra, Indonesia (29) which formed in late Pleistocene time as a result of the eruption of some 500 cubic miles of pumice, deposited on a plateau surrounding the lake. The depression is about 60 miles long by 20 miles wide. Another volcano-tectonic depression of complex origin encompasses Lake Taupo and Lake Rotorua in North Island, New Zealand. Subsidence was in large part caused by emission during Pliocene time of some 2,000 cubic miles of incandescent pumiceous sand that underlies the basin and also builds the surrounding plateaus. Many tectonic features of the depression are due to much later eruptions, and subsidence along faults near Lake Taupo continues to the present day.

7. *Volcano Products: Lavas:* The main products of volcanoes are lavas, ashes or tuffs, and exhalations of gas and vapor. Viscous silicic lava tends to form cumulo-domes or coulees or explodes to produce pyroclastic materials; whereas more fluid lava flows and congeals to form lava flows. Flow

of lava is laminar and not turbulent (30). Lava rivers (Plates 106, 107) can attain velocities of up to 40 miles per hour on steep slopes, but the viscosity of even highly fluid basaltic lava at 1,100°–1200° C is about a million times greater than that of water at 20° C. Measured viscosities (31) of basaltic lava are: 300 poises at 1,400° C, 3,000 poises at 1,200° C, 30,000 poises at 1,150° C (onset of crystallization). Lavas with higher silica content are more viscous. Measured temperatures of mafic lavas are normally in the range of 1,000°–1,200° C; more silicic lavas have temperatures of 800°–1,000° C. Fluid lava spreads out as thin sheets, piling up to form lava plateaus. Generally such lavas are basaltic but phonolite sheets also can form lava plateaus, as they do in Kenya (Plate 90A). Phonolite is an alkaline rock consisting of alkali feldspars and feldpathoids, with sodic pyroxenes and amphiboles.

According to their surface features, lava flows are classified as pahoehoe, aa, and block flows (32) (Fig. 8). *Pahoehoe* or *ropy lava* (Plates 141–148) is mostly fluid and spreads as thin sheets with smooth, glistening surfaces, commonly twisted into ropy wrinkles through continued flow of lava below the elastic skin. *Aa flows* (Plates 140–141) generally move more slowly, at rates of less than 5 feet per minute, and their surface consists of jagged stony clinkers with numerous projections bristling with sharp points and angles (Plate 137). Pahoehoe flows have been observed to change into aa flows farther from their source. This observation lends credence to the theory (33) that pahoehoe lava is rich in volatiles, whereas aa lava is formed during or after expulsion of volatiles from the lava. Continued flow of pahoehoe lava may break the twisted ropy surface to extremely rough, jagged blocks. Such blocky pahoehoe flows (Plate 144B) resemble aa flows in roughness of surface but lack the characteristic aa spines and sharp projections. Craters of the Moon National Monument, Idaho provides examples of all three types of flows but blocky pahoehoe flows are most common. *Block lavas* (Plate 139) are generally more silicic and viscous. The angular blocks have sharp edges and even, plane surfaces without the sharp protrusions and spines of aa lavas. Block lavas also advance very slowly and, like aa lavas, do so with a steep, high front (Plate 89B) from which blocks continually spall to show parts of the glowing molten interior of the flow. Some block lavas may not be lava flows at all but block-and-ash slides or *ladus* (Plate 102) of pyroclastic origin.

Basaltic lava that has poured out under water commonly forms *pillow lavas* (Fig. 8D) (Plates 155–156), piles of ellipsoidal filled sacks, generally 3–4 feet in diameter. Like pahoehoe lavas, pillow lavas owe their origin to the property of basaltic magma to form a tough, elastic skin that contrasts in texture with the stony lava filling the pillows (34). The sacks indent one another and are commonly elongated in the direction of flow. At the lower surface of the pillows there may be V-shaped protrusions that fit between the underlying sacks. Spaces between the pillows are commonly filled with nonvolcanic marine muds. Pillows frequently show radial columnar shrinkage joints (Plate 155B). Because the sacks have convex upper surfaces and flattened or irregular lower surfaces, pillow lavas are of value in determining attitudes of folded beds (35).

Other features found on lava flows (36) include small *spatter* and *cinder cones* (Plates 123–127); *tumuli (Schollendomes)* (Plate 129) and *squeeze-ups* (Plate 136); *lava blisters* (Plates 134–135), small domes raised by pockets of steam; and *pressure ridges* (Plates 130–131), up to 50 feet high and commonly several hundred feet long, generally

with a medial crack along the crest of the ridge. *Lava caves* (Plate 120A) and *tunnels* (Plates 115–116), with or without lava stalactites and stalagmites (Plate 138), are formed when liquid lava drains from within its casing of solidified lava as the supply from the vent diminishes. Foundering of the crusts of such lava streams results in *collapse depressions* (Plates 121–122). Thick flows may show several *flow units* (37) (Plate 88B), in which lava has remained liquid and continued to flow for longer or shorter durations. Some flows show prominent *columnar jointing* (Plates 150–154); others do not (38). Columns grow at right angles to isothermal surfaces within the cooling lava. If the flow is thin or if cooling is rapid and irregular, columns may not form; instead the lava breaks into hackly fragments or joints irregularly. Many flows show two-tier columnar jointing, a lower tier of thick, well-shaped columns and an upper tier of thin and more irregular columns. The two meet along a prominent parting about two-thirds of the distance from the top of the flow. In normal static crystallization of a flow, isothermal surfaces are nearly horizontal and the two tiers stand vertical. If flow of lava is resumed after the columns have grown some distance from the top and bottom of the flow, isothermal surfaces are tilted, hence in the center of the flow, inclined columns form that indicate flow direction. Filled lava tubes may show characteristic war bonnet jointing (Plate 153). Near the fronts of some thick flows there are longitudinal, vertical tension fractures, restricted to the upper parts of the flows. These cracks act as cooling surfaces, deflecting columnar joints towards them.

Lavas may be glassy (*holohyaline*), partly glassy (*hypohyaline*), or stony (*holocrystalline*). Commonly, upper and lower surfaces of lava flows, if not glassy, are finer grained than the interior parts. The upper chilled zone is generally thicker than the lower chilled zone. The top surface may be oxidized and altered to a bright red, friable or clayey material, sometimes called *bole*, while near the base of a flow there may be small inclusions of any underlying flow. Glassy lava commonly shows *flow lines* (Plates 165–166).

High-silica glasses are called *obsidian* (Plate 166B) when black and lustrous; *pitchstone* when dull and waxy. The appearance also reflects the water content: pitchstones have up to 10% H_2O, *perlites* (Plate 164A) have 2–5% H_2O and include dull, waxy glasses as well as black, lustrous glasses. Obsidians have less than 2% H_2O. Silicic glasses are more common than basaltic glass (called *sideromelane* or *tachylite*) (39), because silicic magmas are highly viscous and therefore exhibit greater tendency to undercool than basaltic magma. Volcanic glasses are generally Triassic or younger in age; older glasses and also many younger glasses have crystallized or *devitrified* to very fine grained crystalline aggregates.

Boiling of water or release of gases in lava may result in the formation of round or ovoid spaces which can be open or filled with such minerals as feldspars, zeolites, chlorites, calcite, chalcedony, agate, etc. When empty and small, they are *vesicles* (Plate 162C-D), the rock being called vesicular. *Pumice* is an extremely vesicular, porous, high-silica lava. The dark-colored basaltic equivalent is called scoria (Plate 162B); the corresponding adjective is scoriaceous. Glassy and partly glassy lavas may have small spherical bodies (*spherulites*) (Plates 163B-D), consisting of radiating slender crystals of feldspars or zeolites; such rocks are spherulitic. Spherulites are primary when flow lines in the glass bend around them; secondary spherulites have formed after quenching of the lava to glass, hence flow lines pass through them. Large, hollow, roselike spherulites are called *lithophysae* (Plate 164B). Mafic and intermediate lavas (SiO_2 <55%) commonly have almond-shaped or irregular filled spaces called amygdules or amygdales (Plate 155C); the corresponding rocks are amygdaloidal. The amount and size of amygdales may increase near the upper and lower surfaces of a flow. Pipe vesicles (Plate 162D) and amygdales, tubular in shape, have been found near the lower surface of some flows and may be bent in the direction of flow of the lava.

Most volcanic piles and lava plateaus include intrusive sheets as well as lava flows but generally it is extremely difficult to distinguish one from the other. Dike rocks may become vesicular or amygdaloidal at depths of 1,000 feet or more from the original surface. The presence or absence of vesicles or amygdales therefore does not distinguish extrusive flows from shallow intrusive sheets. The presence of an unchilled, rolling upper surface positively identifies a flow but its absence does not identify an intrusive sheet. Probably the only positive criterion for an intrusive sheet within a pile of lava flows is the presence of inclusions of an overlying flow near the top of the sheet, a somewhat rare event.

Lavas may be *porphyritic* (Plate 163B), i.e., contain large crystals or *phenocrysts* of light-colored (felsic) or dark-colored (mafic) minerals, set in a fine-grained groundmass. Phenocrysts have formed by *intratelluric crystallization* at depth before extrusion of the lava and they may show evidence of their complex history in *reaction rims* or in *resorption* effects that have resulted in strong embayments of the crystals. Phenocrysts may also be aligned parallel to the direction of flow.

Inclusions (*enclaves*) in lava flows may consist of foreign or cognate materials. Foreign inclusions are called *xenoliths* (foreign single crystals are xenocrysts), and these may yield valuable clues in determining the nature of the country rocks under a volcanic pile (Plate 169A) or give information as to the temperature of the magma (Plate 169B). Cognate inclusions are *autoliths* (Plate 169C). The most common autoliths in basaltic lavas are olivine-rich. Mineral compositions in these ultramafic autoliths are remarkably uniform the world over (40), suggesting that the autoliths represent fragments of the upper mantle caught up in the basaltic magmas.

8. *Volcanic Products: Pyroclastic Ejecta* or *Tephra*: The term tephra (41) includes all clastic products of volcanoes. Such fragmentary materials are divided according to grain size (42) into *dust* (<¼ mm diameter), *ash* or *sand* (¼–4 mm diam.), *lapilli* 4–32 mm diam.), and *bombs* or *blocks* which exceed 32 mm (1¼ inch) in diameter. Bombs (Plate 161) are rarely angular, more commonly ellipsoidal, globular, or pear-shaped, partly as a result of rotation in the air during flight. *Breadcrust bombs* (Plate 161A, B) have fissured surfaces due to expansion of the interior after solidification of the crust. Lithified deposits of the finer materials are *tuffs* (Plate 168); those consisting primarily of coarser fragments are agglomerates (Plate 171) or volcanic breccias (Plate 172–174). *Pépérites* (43) are intrusive breccias produced by intrusion of magma into wet sediments. *Crystal* or *lithic tuffs* consist largely of angular fragments of crystals. Characteristics of tuffs are: high porosity, sharply broken crystals, glass shards (which may be devitrified) with cuspate shapes, multimineralic fragments, variable grain size, and crude to well-developed graded bedding and stratification. Some tuffs contain so-called *incompatible minerals* such as quartz and olivine, quartz and nepheline, or quartz and leucite. Bedding and sorting of subaqueous tuffs are generally more regular than of subaerial tuffs (Plate 175B–D). Beneath bombs enclosed in fine tuffs, *sag structures* (Plate 160B) are commonly developed owing to impact and sub-

sequent burial. Many small faults offset the beds and result from settling and compaction of the tuffs (Plate 177). Well-preserved animal or plant casts are not uncommon and many tuffs contain appreciable amounts of carbon. Upon deposition, ashes are normally porous and friable; they are easily eroded unless consolidated (Plate 180), most commonly by silicification through the agency of circulating ground water or water released upon compaction. *Bentonites* (44) are clayey rocks formed by devitrification and alteration of high silica tuffs. They swell readily or crumble into granular aggregates. The main mineral in bentonite is montmorillonite. Some palagonite tuffs (Plates 173A, 174D) are altered basaltic tuffs that contain *palagonite* (45), a yellow or orange, isotropic amorphous material formed by alteration of basaltic glass. Other palagonite tuffs are formed by absorption of water vapor by basaltic glass shards while still hot.

Volcanic clouds (Plates 7, 8) from which ashes are deposited are normally cool enough for the particles in such clouds to be solid. The ashes are distributed radially around the volcano if explosive discharge was vertical (Plate 3B), or are concentrated in a particular sector if discharge was lateral (Plate 2B), or the pattern of the distribution may reflect the prevailing wind direction (Plate 8A). The 1912 eruption in the Valley of Ten Thousand Smokes, Alaska, resulted in ejection of some five cubic miles of pumice and left a Krakatau-type caldera more than two miles across in the summit of Mount Katmai (Plate 101). Prevailing winds came from the northwest; at Kodiak, 100 miles to the southeast, volcanic ashes accumulated to a depth of one foot and the weight of ash deposited on roofs caused the collapse of several homes. Fallout is determined primarily by gravity, large fragments being concentrated near the crater whereas fine dust is carried far downwind (46).

Other volcanic ash clouds are incandescent; these are the dreaded *nuées ardentes* that at their base have glowing avalanches of unsorted tephra or incandescent ash flows. Nuées ardentes move silently and have great mobility. They owe their mobility to continous emission of gas from semiliquid particles and also in part to extreme internal turbulence caused by envelopment of cold air by the hot, debris-laden avalanche (47). Air entrapped in the cloud is heated by the particles, expands explosively, and the fragmented debris of the avalanche is thereby cushioned and buoyed up. This condition eliminates friction between the fragments and the emulsion of gas and solid or semiliquid particles has the lack of coherence and readiness to flow that characterize liquids. Nuées ardentes are released by vertical or lateral discharge from a crater or from the flanks of tholoids (Plate 62). The disastrous 1902 eruption (48) of Mont Pelée, Martinique resulted in lateral discharge of a nuée ardente (Plate 3A) from the side of the dome that had formed in the crater, just opposite a deep notch in the crater rim. The glowing avalanche, with its characteristic dense black cloud, rushed silently down the flank of the volcano, jumped the Riviere Blanche (then a deep ravine), and overwhelmed the town of St. Pierre, leaving a prisoner in the local jail as the only survivor of the 28,000 inhabitants. There was no grading of any sort in this avalanche and large blocks, lapilli, sand, and fine dust were carried en masse for long distances, propelled in part by gravity and the initial explosive force, in part by the self-explosive properties of the fragments during transit. The temperature of the avalanche at St. Pierre was not high enough to melt copper (1058°C) but sufficiently high to melt bottle glass (650°–700°C). The same year the Soufriére volcano (49) of St. Vincent, also in the West Indies, erupted a vertically discharged nuée ardente from an open crater 2,500 feet deep. The boiling emulsion frothed over the crater rim and rushed down the flanks of the volcano in all directions except the northern side, where stood the remains of a large older crater wall. The deposits here were radially distributed and consisted mainly of fine materials; larger fragments and blocks fell back into the crater or its immediate vicinity.

Another example (Fig. 9) has become well-known through the publications of C. N. Fenner (50), although there were no human witnesses to the 1912 eruptions in the Valley of Ten Thousand Smokes, Alaska. A series of eruptions from a newly formed vent, Novarupta (Plate 62B) and from numerous minor explosion vents and fissures produced a rapid succession of nuées ardentes; incandescent sand flows that spread and flooded the valley formed a deposit of sandy tuff 53 square miles in extent and over 700 feet thick in places (Plate 101B). The nuées ardentes were hot enough to carbonize or burn trees and brush near the lower end of the valley, which has been called Ten Thousand Smokes because of the numerous fumaroles that continued activity on the flat featureless surface of the tuff deposit for many years after the eruptions.

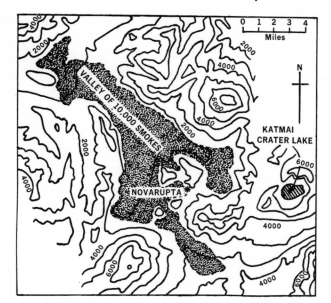

Fig. 9: Valley of Ten Thousand Smokes, Alaska (Modified from Fenner, 1923)

Deposits of nuées ardentes may be consolidated or unconsolidated. In general, bedding is poorly developed or absent and sorting is also poor. The deposits are commonly indurated by vapor-phase crystallization. Such compacted, indurated deposits of nuée ardente origin are called *sillar* (51) (Plates 167, 168). The sand deposit in the upper part of the Valley of Ten Thousand Smokes is indurated and firmly consolidated by tiny secondary growths of tridymite and orthoclase formed shortly after deposition. However, in the lower part of the valley where temperatures may have been less, the deposit has little coherence. Lateral discharge of nuées ardentes commonly gives rise to unsorted deposits in which there are sharply angular fragments of all sizes (Plate 102A). *Ladus* are glowing *block-and-ash slides* in which large blocks predominate (Plate 102B). They may form by relatively small nuées ardentes or by spalling of the fronts of slowly advancing coulees on the steep upper slopes of a volcano.

If temperatures of ash flows are sufficiently high and particles are plastic upon deposition, the glass fragments can agglutinate under the weight of overlying materials to form *welded tuffs* (52). Welding occurs at temperatures in excess of 575°C and probably as high as 900°C in some ash flows. Single ash flows are welded and cool to simple units with a basic pattern of zones (53). The interior of the unit has a zone of dense welding; above and below it are successively the zone of partial welding and of no welding, both being thicker above than below the zone of dense welding. Crystallization produces a superimposed zonal pattern; types of crystallization include devitrification in the zones of dense and partial welding, vapor-phase crystallization in the porous parts of the unit, and fumarolic alteration in the upper zone of no welding. Very thick units can also include a zone of granophyric crystallization within the zone of dense welding. Variations in these basic zonal patterns are numerous and are produced by differences in original temperatures and thicknesses of ash flows, buried topography, rapid succession of ash flows producing compound units, induration or removal of upper parts of units by later ash flows, removal of upper parts of units by erosion, or much later silicification and recrystallization. Collapse, flattening, and alignment of pumice fragments in the lower parts of welded tuff deposits commonly result in a crude layering or foliation (Plate 168). The consolidated rocks ring to the hammer and may show well-developed columnar jointing (Plate 152B) owing to contraction upon cooling or to tectonic stresses.

In the past, welded tuffs were mistaken for extensive flows of high-silica lavas (rhyolites). However, a viscid lava flow has a scoriaceous surface and a brecciated base formed by inclusion of blocks and talus broken off the steep front of the flow and overrun by the advancing lava. Both these characteristics are lacking in welded tuffs. The upper surface of a high-silica lava is strongly convex whereas that of a welded tuff is more nearly horizontal. It is evident that deposits of sillar and of welded tuff cannot always be distinguished. The term *ignimbrite* (54), as now used, covers both types of ash-flow tuffs. Many examples of extensive ignimbrite sheets have been described. The Bishop sheet (55) of southeastern California covers an area of 400 square miles and is 500 feet thick. Chiricahua National Monument, Arizona (56) has excellent exposures of a succession of nine welded tuff units. The supposed rhyolites of the Yellowstone National Park, Wyoming are predominantly ignimbrite sheets (57) and cover a total area of 4,000 square miles. Early Tertiary ignimbrites in the Great Basin cover about 80,000 square miles, locally to a thickness of 8,000 feet. The volume of the Taupo-Rotorua ignimbrite sheets of North Island, New Zealand is estimated at about 2,000 cubic miles (58). Thus, ignimbrite sheets form the pyroclastic high-silica counterparts of plateau-building basaltic lavas, and both result from fissure eruptions. Note, however, that no eruption of ash flows that produced welded tuffs has ever been observed.

Mudflows may be volcanic or nonvolcanic in origin. Nonvolcanic mudflows effect the redistribution of the porous pyroclastic materials deposited on and around a volcanic cone. Volcanic mudflows or *lahars* (Plates 98–99) result from mingling of nuées ardentes with river waters, melting of glacial ice by lava flows or nuées ardentes, or from ejection of crater lakes at the outbreak of volcanic eruptions (59). The Miocene-Pliocene Mehrten formation of the Sierra Nevada is mainly of lahar origin. The formation is predominantly andesitic, up to 4,500 feet thick, and has a volume of some 2,000 cubic miles. Many destructive lahars of Merapi volcano in Java, Indonesia can be traced to mingling of nuées ardentes with river waters., In the 1877 eruption of Cotopaxi, Ecuador (60) nuées ardentes cut channels 100–200 feet deep into the glaciers of the volcano and caused a large number of mudflows that flooded the surrounding plains. The 1919 eruption of the Kelut volcano in Java, Indonesia evacuated the crater lake which contained some 30 million cubic yards of water. The resulting lahar travelled down the flanks of the volcano and caused the loss of 5,000 lives. To prevent a recurrence, a tunnel was drilled through the crater wall which keeps the lake at a constant level. Subglacial eruptions of the Icelandic Grimsvotn volcano in the Vatnajökull (jökull=glacier) in 1934 and 1938 produced large mudflows with per-minute discharges from 5 to 20 times that of the Amazon River. Lahars form *volcanic conglomerates* (Plate 174C), chaotic assemblages of large and small, rounded to subangular blocks in a matrix of gravel, sand, or mud. Bedding is poorly developed or absent in these deposits.

Igneous breccias are subdivided into three main types (61); autoclastic, pyroclastic, and epiclastic. *Autoclastic breccias* (Plate 173C) are produced by magma movements, and include flow breccia and various intrusion breccias. *Pyroclastic breccias* (Plate 173B) are produced by gas explosions. *Epiclastic breccias* are caused by any other process and are transported by any epigene geomorphic agent.

Reference must also be made to peculiar intrusive fragmented rocks of complex origin which closely resemble volcanic pyroclastic products. The fine-grained rocks are called *tuffisites* whereas coarse conglomeratic or agglomeratic rocks of this type are known as *explosion breccias* to distinguish them from volcanic breccias and from pépérites (62). Tuffisites and explosion breccias apparently are formed through explosive disruption of country

rocks by gas and fluidization of the disaggregated products (63). British examples have been described from Slieve Gullion, Tory Island, and the Isle or Rhum (64). In the Absaroka volcanic field, Wyoming (Plate 171D, 172D), there are intrusive and extrusive breccias of similar origin (65), with all gradations from autoclastic and pyroclastic breccias formed beneath the surface to epiclastic volcano breccias of lahar origin. Note that autoclastic extrusive or intrusive breccias are formed by relatively quiet processes of autobrecciation, as distinct from pyroclastic breccias.

9. *Volcanic Products*: *Exhalations*: The violence of volcanic eruptions precludes collection of gases escaping during the most active phases of volcanic activity. However, many measurements have been made of the magmatic gases expelled from ponded lava and from Halemaumau Lava Lake on Kilauea, Hawaii before destruction of the lake in 1924. Gases from fumaroles have also been analyzed, as well as gases released by volcanic rocks upon melting *in vacua* in the laboratory. Results (66) are notably variable but variations are largely accounted for by shifts in gas equilibria at different temperatures (67). A typical composition of Hawaiian volcanic gases is: $H_2O=79.31$; $CO_2=11.61$; $SO_2=6.48$; $N_2=1.29$; $H_2=0.58$; $CO=0.37$; $S_2=0.24$; $Cl_2=0.05$; and $Ar=0.04$ (volume percent) (68). Water is the chief constituent of volcanic gases, amounting in most cases to 70–95% of the total. The amounts of gas released are enormous; it is estimated that a cubic yard of new lava at 1,200°C might release more than 50 cubic yards of gas at a pressure of 50 atmospheres. Probably much of the water in volcanic exhalations is groundwater absorbed by the rising magma or by hot volcanic glasses. The contribution of water of direct magmatic origin in volcanic thermal springs is small, probably less than 5% (69). Even after extrusion, obsidians may become hydrated to pitchstones or perlites (70) and thereby raise their water content from 0.3% or less to several (weight) percent. Gas combustion has been advocated by earlier writers as one of the principal sources of volcanic heat but this is now largely discredited (71). The question of the heat supply that maintains the volcanic furnace is still a major unsolved problem in volcanology.

There is better understanding of the explosive frothing of lava which causes the destructive nuées ardentes (72). Morey has discussed the analogy of the cooling of a melt of K_2SiO_3 saturated with water at 1 atmosphere. At the melting point, the liquid takes up about 1% H_2O which lowers its melting point by 35°C. Upon quick cooling, the melt becomes supersaturated. First a few bubbles appear in the under-cooled melt; then suddenly bubble formation becomes rapid and the viscous melt swells into a pumiceous mass many times its original volume. This is an example of the so-called *second boiling* at atmospheric pressure. Shepherd and Merwin (73) heated obsidian that was found to contain 55 cc of gas per gram of glass at 1,200°C (0.9% by weight). Upon heating in a vacuum, the glass frothed to a pumiceous mass at 850°C. After the gas had escaped, the fine threads of glass in the pumice became quite rigid and had to be heated to above 1,250°C before they melted. Verhoogen concludes that the rate of vesiculation is probably the most important single factor in determining the eruptive behavior of magma. Swift rise of gas bubbles causes lava to boil, as it does in lava lakes. If bubbles expand more rapidly than they rise, their coalescence can result in violent disruption of the lava. If the gas pressure of the bubbles is high at that time, there may be a violent outburst with ejection of volcanic ashes; if the gas pressure is low, nuées ardentes may result. Many complex factors are involved: the viscosity of the magma, rate of pressure release, and incubation effects which can cause nucleation of bubbles to lag behind over-saturation. Effervescence, once begun, tends to proceed at an accelerating pace.

10. *Quantitative Estimates of Volcanic Products:* Estimates of the quantitative output of individual volcanoes are both poor and incomplete. Parícutin, Mexico is the only volcano for which there is a complete and reasonably reliable record (74). The volcano started in a cultivated field on February 20, 1943 and in nine years built a cone nearly 1,300 feet high (Plate 1A–B). All activity ceased on March 4, 1952.

Total amounts of pyroclastic material and lava erupted from early 1943 to early 1952 are 2230 million metric tons and 1330 million metric tons respectively.

The amount of water vapor emitted in 1945 was calculated at 18,220 tons per day which reduces to a water content of 1.1% (by weight) in the magma (the solidified lavas of Parícutin contain an average of 0.1% H_2O). Assuming that the proportion of ejected solids to water vapor remained constant throughout the life of the volcano, this gives a total output of 52 million tons of water, the equivalent of a body of water 3¾ miles square and 1 yard deep.

There are also (less reliable) estimates (75) for the total volcanic output over the world for the last four centuries. The totals are: 83 cubic miles (320 cubic km.) of pyroclastic materials and 15½ cubic miles (50 cubic km.) of lava, or an average total output of about 0.3 cubic miles per year. The central Pacific shield volcanoes produced in this period 3 cubic miles of lava and a half cubic mile of pyroclasts; the Indonesian volcanoes of the Sunda belt produced one-eighth of a cubic mile of lava and 46 cubic miles of volcanic ashes.

Although this figure of 83 cubic miles of tephra may well be over-estimated, there is an undoubted preponderance of fragmental materials in the volcanic output of the world. Because deposits of pyroclastic materials are rapidly eroded and re-distributed, it is easy to underestimate the amounts of silicic volcanic products, especially those formed in the geologic past. Earlier evaluations of areas and volumes of fragmental volcanic materials are almost certainly too low. Bentonite beds (76) cover areas in the United States east of the Rocky Mountains totalling 300,000 to 400,000 square miles, indicating a volume of some 4,000 cubic miles. Farther west, these areas and amounts are even larger. The volume of ignimbrites in the Great Basin is probably more than 50,000 cubic miles (77). Rhyolitic rocks of the Potosi volcanic series of the San Juan Mountains, Colorado have a volume of 2,300 cubic miles (78), equivalent to a granite pluton 230 square miles in area and 10 miles deep. Silicic volcanic ashes are probably the main igneous sources of sediments.

11. *Sills, Dikes, and Laccoliths: Sills* are concordant, tabular injected bodies, thin compared with their areal extent (Plates 78–79). The actual thickness ranges from a few feet to a thousand feet or more. The Palisade sill, New Jersey is apparently a simple, differentiated dolerite sill with a maximum thickness of 1,000 feet (Plate 79B). It is the intrusive equivalent of the Triassic flows of Watchung basalt that outcrop farther west. Another well-known dolerite sill is the Whin sill of northern England (Plate 79A). Its thickness averages only 100 feet but the length of its outcrop trend is over 80 miles and the extent of the sill underground is more than 1,550 square miles. In parts of the Union of South Africa, Karroo formations have been extensively invaded by dolerite sills; in some areas the proportion of igneous to sedimentary rocks reaches 1:3. Sills may consist of most types of igneous rocks but dolerite sills are most common. Like other injected bodies, sills generally have

well-defined, fine-grained chilled contacts with closely spaced contraction joints. The fine-grained contact phases represent portions of the magma quenched immediately after emplacement. The contact rocks are commonly porphyritic; the phenocrysts indicate the amounts and types of minerals that crystallized before emplacement of the magma. The chilled contacts grade into the coarser-grained rocks of the interior of the sill, in which the jointing is generally more widely spaced than in the contact zones. Strips of the country rocks may be wedged or stoped from the upper or lower contact to form xenoliths within the sill. Less commonly, the chilled contact itself has been stripped off and is found as autoliths of fine-grained rocks some distance from the contact. Near the contacts, the country rocks are indurated and baked or have recrystallized to hornfelses by contact metamorphism. Although apparently concordant, many intrusions are not true sills but transgress the invaded formations in series of small steps or at small angles ($<20°$). Such slightly transgressive tabular bodies are called *sheets* (Plate 80). Sheets also include essentially horizontal intrusions emplaced in massive igneous rocks such as granites.

Dikes are discordant, tabular injected bodies, thin compared with their length. Actual dimensions may range from a fraction of an inch wide and less than a foot long to widths of several hundred feet and lengths of scores of miles. Small ancillary dikes may run parallel to a major dike or extend as apophyses or tongues from it. Dike systems that are more or less parallel or that radiate from a common center are called *dike swarms* (Plate 78A). Dike outcrops are normally straight and continuous but irregular, discontinuous dikes are also known. Generally, they have no upward or downward terminations but a few examples of *roofed dikes* have been described, the best known being the 130 mile long Cleveland dike of northern England. Dikes are emplaced along fractures opened by crustal tension and they are also conduits for magma ascending through the earth's crust. They may feed other injected bodies or terminate in lava flows. Large amounts of magma or lava can pass through relatively thin *feeder dikes* (Plate 77B). Contact metamorphic effects cannot always be linked to the thickness of dikes. A thick dike emplaced by a single surge of magma may

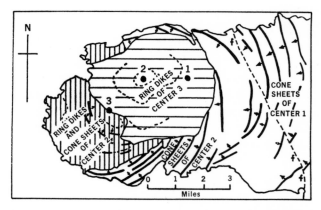

Fig. 10:   Cone Sheets and Irruptive Centers of Ardnamurchan, Scotland [Modified from Richey and Thomas, 1930, Pl. 2, p. 71 (81)] Horizontal Shading: Ring Dikes of Center 2; Vertical Shading: Ring Dikes of Center 3.

have only a thin baked zone along its walls whereas a thin feeder dike through which magma flowed for a protracted time may be associated with zones of intense contact metamorphism.

Ring dikes have arcuate outcrops and are vertical or inclined outward. The type example is the Loch Ba ring dike on the island of Mull, Scotland. The Ossipee Mountains ring complex (79) in New Hampshire has a nearly perfect subcircular ring dike. *Cone sheets* (Plate 80B) also have arcuate outcrops but are less steep than ring dikes and incline inward. Cone sheets and vertical ring dikes result from upward surge of magma; ring dikes inclined outward are formed by withdrawal of magmatic support (80). Numerous ring dikes (Fig. 10) and conesheets are associated with the irruptive centers of Ardnamurchan, Scotland (81) (Plate 37A). Ring dikes and cone sheets commonly surround central stocks, forming ring complexes produced by cauldron subsidence. Many such ring complexes have been described from New Hampshire (Plate 37B), the Oslo district of Norway, and Nigeria (Plate 67B) (82).

Laccoliths are concordant intrusions that have arched up their roofs; they have the shape of planoconvex or doubly convex lenses (Plate 68). Initial doming in laccoliths is probably due to upsurge of magma upon emplacement but doming may be accentuated by subsequent marginal cooling and the increasing viscosity of the magma (83). Fluid magmas tend to spread laterally to form sills or sheets, even if initially there is some arching of the roof.

REFERENCES

(1) General: Rittmann, A. 1962. *Volcanoes and their Activity*. New York: Interscience Publ. 305 pp.; Bullard, F. M. 1962. *Volcanoes in History, in Theory, in Eruption*. Univ. of Texas Press, 441 pp.; Cotton, C. A. 1952. *Volcanoes as Landscape Forms* 2nd ed. Christchurch, New Zeal: Whitcombe and Tombs, 416 pp.; Zies, E. B. 1938. Surface Manifestations of Volcanic Activity. *Trans. Am. Geophys. Union*, 19: 10–38; MacDonald, G. A. 1968. Forms and Structures of Extrusive Basaltic Rocks: In *Basalts* (Hess and Poldervaart, editors) New York: Interscience, 1, 1–61.

(2) Sapper, K. 1931. Volcanology, Physics of the Earth. *National Research Council Bull. 77*.

(3) Fuller, R. E. 1927. The Closing Phase of a Fissure Eruption. *Am. Jour. Sci.* 14: 228–230.

(4) Tyrrell, G. W., 1937. Flood Basalts and Fissure Eruptions. *Bull. Volcanolog.* 1: 89–111.

(5) Einarsson, T. 1960. The Plateau Basalt Areas of Iceland. *21st Internat. Geol. Cong., Guidebook*, Exc. A2, Reykjavik, pp. 5–20; Thorarinsson, S. 1960. The Post-Glacial Volcanism. On the Geology and Geophysics of Iceland. *21st Internat. Geol. Cong. Guidebook*. Exc. A2, Reykjavik, pp. 33–45; Barth, T.F.W, 1950, Volcanic Geology, Hot Springs, and Geysers of Iceland: *Carnegie Inst. Wash Publ. 587*.

(6) Stearns, H. T. 1924, Craters of the Moon National Monument, Idaho: *Geog. Rev.* 14, 3: 362–372.

(7) Powers, S. 1916. Volcanic Domes in the Pacific: *Am. Jour. Sci.* 42: 266–67.

(8) MacDonald, G. A. and Hubbard, D. H. 1951. Volcanoes of Hawaii National Park. *Hawaii Nature Notes*, 4, 2: 15; Schultz, P. E. 1943, Some Characteristics of the Summit Eruption of Mauna Loa, Hawaii in 1940. *Bull. Geol. Soc. Am.* 54: 739–746; Stearns, H. T. 1966. *Geology of the State of Hawaii*. Palo Alto, Calif.; Pacific Books, Inc.

(9) Stearns, H. T. 1942. Origin of Haleakala Crater, Island of Maui, Hawaii: *Bull. Geol. Soc. Am.* 53: 1–14.

(10) Bullard, F. M. 1947. The Story of Parícutin Volcano, Michoacán, Mexico. *Bull. Geol. Soc. Am.* 58: 433–450.

(11) Perret, F. A. 1924. The Vesuvius Eruption of 1906. *Carnegie Inst. of Wash. Publ. 339*.

(12) Williams, H. 1932. The History and Character of Volcanic Domes: Univ. California Publ. *Dept. Geol. Sci. Bull.* 21, 5: 51–146.

(13) Lacroix, A. 1904. *La Montagne Pelée et ses eruptions*. Paris: Masson; Lacroix, A. 1908. *La Montagne Pelée apres ses eruptions*: Paris: Acad. Sci. pp. 79–93.

(14) Tanakadate, H. 1930. Two Types of Volcanic Dome in Japan. *4th Pacific Sci. Cong. Proc.*, Batavia, 2B: 695–704.

(15) Day, A. L. and Allen, E. T. 1925. *The Volcanic Activity and Hot Springs of Lassen Peak*. Carnegie Inst. of Washington, 190 pp.; Williams, H. 1932. The Geology of the Lassen Volcanic National Park, California. Univ. California Publ., *Dept. Geol. Sci. Bull.*, 21, 8: 195–385; Putnam, W. C. 1938. The Mono Craters, California. *Geog. Rev.* 28, 1: 68–83; Kistler, R. W. 1966. *Geological Map of the Mono Craters Quadrangle, Mono and Tuolumne Counties, California*. U.S. Geol. Surv., GQ-462.

(16) Williams, H. 1936. Pliocene Volcanoes of the Navajo-Hopi Country. *Bull. Geol. Soc. Am.* 47: 111–171; Shoemaker, E. M., Roach, C. H. and Byers, F. M., Jr. 1962. Diatremes and Uranium Deposits in the Hopi Buttes, Arizona. In *Petrologic Studies, A. Volume to Honor A. F. Buddington*. Geol. Soc. Am. 327–355; McBirney, A. R. 1959. Factors Governing Emplacement of Volcanic Necks. *Am. Jour. Sci.* 257: 431–448.

(17) Williams, H. 1941. Calderas and their Origin: Univ. California Publ., *Dept. Geol. Sci. Bull*, 25, 6: 239–346; Branch, C. 1966. Volcanic Cauldrons, Ring Complexes, and Associated Granites of the Georgetown Inlier, Queensland. *Australia Bur. Mineral Resources, Geol. and Geophys. Bull. 76*, 158 pp.; Johnson, W. 1966. Mechanisms of Cauldron Subsidence: *Nature* 210, 5033: 291–292; MacDonald, G. A. 1965. Hawaiian Calderas. *Pacific Sci.* 19, 3: 141–155; Smith, R. L. and Bailey, R. A. 1968. Resurgent Cauldrons: In

Studies in Volcanology, Geol. Soc. Am. Memoir 116: 613–662; McCall, G. J. H. 1963. Classification of Calderas—Krakatoan and Glencoe Types. *Nature* 197: 136–138.

(18) Furneaux, R. 1964. *Krakatoa*. Englewood Cliffs, N. J.: Prentice-Hall, Inc. 224 pp.

(19) Williams, H. 1942. The Geology of Crater Lake National Park, Oregon, with a Reconnaissance of the Cascade Range southward to Mount Shasta. *Carnegie Inst. Washington, Publ. 540*, 62 pp.

(20) Clough, C. T., Maufe, H. B. and Bailey, E. B. 1909. The Cauldron Subsidence of Glencoe and the Associated Igneous Phenomena: *Geol. Soc. London, Quart. Jour.* 65: 611–676.

(21) Stearns, H. T. and MacDonald, G. A. 1946. Geology and Ground-water Resources of the Island of Hawaii. *Hawaii Div. Hydrography Bull. 9*, 363 pp.

(22) Williams, H. 1935. The Newberry Volcano of Central Oregon. *Bull Geol. Soc. Am.* 46: 253–304.

(23) Kuno, H. 1953. Formation of Calderas and Magmatic Evolution. *Am. Geophys. Union Trans.* 34, 2: 270–272.

(24) Smith, R. L., Bailey, R. A. and Ross, C. S. 1961. Structural Evolution of the Valles Caldera, New Mexico, and its Bearing on the Emplacement of Ring Dikes. *U.S. Geol. Surv. Prof. Paper 424 D*, pp. D145–D149.

(25) Sekiya, S. and Kikuchi, Y. 1890. The Eruption of Bandai-san. *Imper. Univ. Tokyo, Jour. Coll. Sci.* 3: 91–172.

(26) Grange, L. I. 1937. The Geology of the Rotorua-Taupo Subdivision, Rotorua and Kaimanawa Divisions: *New Zealand Geol. Surv. Bull.* 37: 138.

(27) Snyder, F. G. and Gerdemann, P. E. 1965. Explosive Igneous Activity Along an Illinois-Missouri-Kansas Axis. *Am. Jour. Sci.* 263: 465–493.

(28) Von Engeln, O. D. 1932. The Ubehebe Craters and Explosion Breccias in Death Valley, California: *Jour. Geol.* 40, 8: 726–734.

(29) Van Bemmelen, R. W. 1930. The Origin of Lake Toba, North Sumatra: *4th Pacific Sci. Cong. Proc.* Batavia 2A: 115–124; Westerveld, J. 1952. Quaternary Volcanism on Sumatra: *Bull. Geol. Soc. Am.* 63: 561–594.

(30) Wentworth, C. K. 1954. The Physical Behavior of Basaltic Lava Flows. *Jour. Geol.* 62, 5: 425–438; MacDonald, G. A. 1963. Physical Properties of Erupting Hawaiian Magmas. *Bull. Geol. Soc. Am.* 74: 1071–1077.

(31) Nichols, R. L. 1939. Viscosity of Lava. *Jour. Geol.* 47, 3: 290–302; Wentworth, C. K., Carson, M. H. and Finch, R. H. 1945. Discussion on the Viscosity of Lava. *Jour. Geol.* 53, 2: 94–104; Palmer, H. S. 1927. A Study of the Viscosity of Lava. *Bull. Hawaiian Volc. Observ.* 15: 1–4.

(32) Jones, A. E. 1943. Classification of Lava Surfaces. *Am. Geophys Union Trans.* 24th Ann. Meeting, pt. 1: 265–268; MacDonald, G. A. 1953. Pahoehoe, Aa, and Block Lava. *Am. Jour. Sci.* 251, 3: 169–191; Wentworth, C. K. and MacDonald, G. A. 1953. Structures and Forms of Basaltic Rocks in Hawaii. *U.S. Geol. Surv. Bull.* 994, 98 pp.

(33) Jagger, T. A. 1920. Seismometric Investigation of the Hawaiian Lava Column. *Seismo. Soc. Am. Bull.* 10, 4: 162–168.

(34) Lewis, J. V. 1914. Origin of Pillow Lavas. *Bull. Geol. Soc. Am.* 25: 646; Snyder, G. L. and Fraser, G. D. 1963. Pillowed Lavas, Pt. I. Intrusive Layered Lava Pods and Pillowed Lavas, Una-laska Island, Alaska and Pt. II, A Review of Selected Literature. *U.S. Geol. Surv. Prof. Papers 454-B* (23 pp.) *and 454-C* (7p).

(35) Shrock, R. R. 1948. *Sequence in Layered Rocks*. New York: McGraw-Hill Book Co., pp. 359–366.

(36) Nichols, R. L. 1946. McCartys Basalt Flow, Valencia County, New Mexico: *Bull. Geol. Soc. Am.* 57: 1049–1086.

(37) Nichols, R. L. 1936. Flow-Units in Basalt. *Jour. Geol.* 44, 5: 617–630.

(38) James, A. V. G. 1920. Factors Producing Columnar Structure in Lavas and Its Occurrence near Melbourne, Australia. *Jour. Geol.* 28, 5: 458–469; Waters, A. C. 1960. Determining Direction of Flow in Basalts. *Am. Jour. Sci.* 258A, Bradley Volume, pp. 350–366; Beard, C. N. 1959. Quantitative Study of Columnar Jointing. *Bull. Geol. Soc. Am.* 70: 379–381; Spry, A. 1962. The Origin of Columnar Jointing, particularly in Basalt Flows. *Geol. Soc. Australia Jour.*, Part 2. 8: 191–216.

(39) Fuller, R. E. 1931. The Aqueous Chilling of Basaltic Lava on the Columbia River Plateau. *Am. Jour. Sci.* 5th Ser. 21, 124: 281–300; Cucuzza, S. 1963. Proposal for a Genetic Classification of Hyaloclastites. *Bull. Volcanolog.* Ser. 2, 25: 315–321.

(40) Ross, C. S., Foster, M. D. and Myers, A. T. 1954. Origin of Dunites and of Olivine-rich Inclusions in Basaltic Rocks. *Am. Mineral.* 39: 693–737.

(41) Thorarinsson, S. 1955. Discussions. *Bull. Volcanolog.* Ser. 2, 16: 12.

(42) Wentworth, C. K. and Williams, H. 1932. The Classification and Terminology of the Pyroclastic Rocks. *Nat. Res. Counc. Bull. 89*, Comm. Sedim. Rept. pp. 19–23.

(43) Lacroix, A. 1930. Remarques sur les Materiaux de Projection des Volcans et sur la Genese des Roches Pyroclastiques qu'ils Constituent. *Soc. Geol. France, Centenaire Jubilaire* 1830–1930, 2: 431–472; MacDonald, G. A. 1939. An Intrusive

Pépérite at San Pedro Hill, California. *Univ. California Publ. Dept. Geol. Sci. Bull* 24, 12: 329–338.

(44) Ross, C. S. 1928. Altered Paleozoic Volcanic Materials and Their Recognition: *Am. Assoc. Petrol. Geol. Bull.* 12, 2: 143–164; Slaughter, M. and Earley, J. W. 1965. Mineralogy and Geological Significance of the Mowry Bentonites. *Geol. Soc. Am. Spec. Paper 83.*

(45) Peacock, M. A. and Fuller, R. E. 1928. Chlorophaeite, Sideromelane, and Palagonite from the Columbia River Plateau. *Am. Mineral.* 13: 360–383; Bonatti, E. 1963. Palagonite, Hyaloclastites, and Alteration of Volcanic Glass in the Ocean. *Bull. Volcanolog.* 28: 3–15.

(46) Knox, J. B. and Short, N. M. 1964. A Diagnostic Model using Ashfall Data to Determine Eruption Characteristics and Atmospheric Conditions during a Major Volcanic Event. *Bull. Volcanolog.* 26: 455–469.

(47) Perret, F. A. 1935. The Eruption of Mont Pelée 1929–1932. *Carnegie Inst. Washington Publ. 458,* 126 pp.; McTaggart, K. C. 1960, The Mobility of Nuées Ardentes. *Am. Jour. Sci.* 258, 5: 369–382; Gorshkov, G. S. 1959. Gigantic Eruption of the Volcano Bezymianny. *Bull. Volcanolog.* Ser. 2, 20: 77–109.

(48) Lacroix, A. 1904. *La Montagne Pelée et ses Eruptions.* Paris: Masson et Cie., 662 pp; MacGregor, A. G. 1952. Eruptive Mechanisms; Mt. Pelèe, the Soufrière of St. Vincent and the Valley of Ten Thousand Smokes. *Bull. Volcanolog.* Ser. 2, 12: 49–74.

(49) Anderson, T. and Flett, J. S. 1903. Report on the Eruptions of the Soufrière, in St. Vincent, in 1902, and on a visit to Montagne Pelée, in Martinique. Part 1. *Roy. Soc. London, Phil. Trans.* A. 200: 353–553; Hay, R. L. 1959. Formation of the Crystal-rich Glowing Avalanche Deposits of St. Vincent, B.W.I.; *Jour. Geol.* 67, 5: 540–562.

(50) Fenner, C. N. 1920. The Katmai region, Alaska, and the Great Eruption of 1912. *Jour. Geol.,* 18, 7. 569–606; 1923. The Origin and Mode of Emplacement of the Great Tuff Deposit of the Valley of Ten Thousand Smokes: *Nat. Geogr. Soc. Contrib. Techn. Papers,* Katmai Series, no. 1, 74 pp; 1950; The Chemical Kinetics of the Katmai Eruption, Parts 1 and 2. *Am. Jour. Sci.* 248, 9: 593–627 and no. 10, pp. 697–725; Keller, A. S. and Reiser, H. N., 1959. Geology of the Mount Katmai area, Alaska. *U.S. Geol. Surv. Bull. 1058-G,* pp. 261–298; Curtis, G. H. 1968. The Stratigraphy of the Ejecta from the 1912 Eruption of Mount Katmai and Novarupta, Alaska. *Geol. Soc. Am. Memoir 116:* 153–210.

(51) Fenner, C. N., 1948. Incandescent Tuff Flows in Southern Peru. *Bull. Geol. Soc. Am.* 59: 879–893.

(52) Ross, C. S. and Smith, R. L. 1960. Ash Flow Tuffs: Their Origin, Geologic Relations, and Identification. *U.S. Geol. Surv. Prof. Paper 366,* 81 pp; Smith, R. L. 1960. Ash Flows. *Geol. Soc. Am. Bull.* 71: 795–842; Yokoyama, L. 1966. Crustal Structures that produce Eruptions of Welded Tuff and Formation of Calderas. *Bull. Volcanolog.* 29; 51–60.

(53) Smith, R. L., 1960, Zones and Zonal Variations in Welded Ash Flows: *U.S. Geol. Surv. Prof. Paper 354F,* 159 pp.

(54) Marshall, P., 1935, Acid rocks of the Taupo-Rotorua Volcanic District: *Roy. Soc. New Zealand Trans. and Proc.* part 3, 64: 323–366.

(55) Gilbert, C. M. 1938. Welded Tuff in Eastern California. *Bull. Geol. Soc. Am.* 49: 1829–1862.

(56) Enlows, H. E. 1955. Welded Tuffs of Chiricahua National Monument, Arizona. *Bull. Geol. Soc. Am.* 66: 1215–1246.

(57) Boyd, F. R. 1961. Welded Tuffs and Flows in the Rhyolite Plateau of Yellowstone Park, Wyoming. *Bull. Geol. Soc. Am.* 72: 387–426.

(58) Marshall, P. 1932. Notes on Some Volcanic Rocks of the North Island of New Zealand. *New Zeal. Jour. Sci. and Tech.* 13: 198–200.

(59) Anderson, C. A. 1933. The Tuscan Formation of Northern California: Univ. California Publ. *Dept. Geol. Sci. Bull.* 23, 7: 215–276; Curtis, G. H. 1954. Mode of Origin of Pyroclastic Debris in the Mehrten Formation of the Sierra Nevada. Univ. California Publ. *Dept. Geol. Sci. Bull.* 29, 9: 453–502.

(60) Wolf, T. 1878. Geognostische Mitteilungen aus Ecuador. 5. Der Cotopaxi und seine letzte Eruption an 26 Juni, 1877. *Neues Jahrb. Mineral. Geol. u. Paleont.* pp. 113–167.

(61) Fisher, R. V. 1960. Classification of Volcanic Breccias. *Bull. Geol. Soc, Am.* 71: 973–982; 1961. Proposed Classification of Volcanoclastic Sediments and Rocks. *Bull. Geol. Soc. Am.* 72: 1409–1414.

(62) Cloos, H. 1941. Bau und Tätigkeit von Tuffschloten; Untersuchungen an dem Schwabischen Vulkan: *Geol. Rundschau* 32: 709–800.

(63) Reynolds, D. L. 1954. Fluidization as a Geological Process, and Its Bearing on the Problem of Intrusive Granites. *Am. Jour. Sci.,* 252, 10: 577–614.

(64) Richey, J. E. 1932. The Tertiary Ring-complex of Slieve Gullion, Ireland. *Geol. Soc. London, Quart. Jour.* 88: 776–849; Reynolds, D. L. 1951. The Geology of Slieve Gullion, Foughill, and Carickcarnan, an Actualistic Interpretation of a Tertiary Gabbro-Granophyre Complex. *Roy. Soc. Edinburgh Trans.* 62: 85–143; Whitten,

E.H.T. 1959. Tuffisites and Mangerite Tuffisites from Tory Island, Ireland, and related Products of Gas Action. *Am. Jour. Sci.* 257, 1: 113–137; Hughes, C. J. 1960. The Southern Mountains Igneous Complex, Isle of Rhum. *Geol. Soc. London Quart. Jour.* 116: 111–138.

(65) Hay, R. L. 1964. Structural Relationships of Tuff Breccia in Absaroka Range, Wyoming. *Bull. Geol. Soc. Am.* 65: 605–620; Parsons, W. H. 1960. Origin of Tertiary Volcanic Breccias, Wyoming. *21st Internat. Geol. Cong. Rept.* Copenhagen, Part 13, pp. 139–146.

(66) Shepherd, E. S. 1948. The Gases in Rocks and some Related Problems. *Am. Jour. Sci.* 5th Ser. Day Volume, 35A: 311–351; Jagger, T. A. 1940. Magmatic Gases. *Am. Jour. Sci.* 238, 5: 313–353.

(67) Ellis, A. J. 1957. Chemical Equilibrium in Magmatic Gases. *Am. Jour. Sci.* 255, 6: 416– Krauskopf, K. B. 1959. The Use of Equilibrium Calculations in Finding the Composition of a Magmatic Gas Phase. In *Researches in Geochemistry*. P. H. Abelson, ed. New York: J. Wiley and Sons, pp. 260–278.

(68) Eaton, J. P. and Murata, K. H. 1960. How Volcanoes Grow. *Science.* 132. 3432: 938; Daly, R. A. 1911. The Nature of Volcanic Action. *Proc. Am. Acad. Arts and Sci.* 47, 3. 47–122.

(69) White, D. E. 1957. Thermal Waters of Volcanic Origin. *Bull. Geol. Soc. Am.* 68: 1643; White, D. E. and Waring, G. A. 1963. Volcanic Emanations, In *Data of Geochemistry*, 6th ed., U.S. Geol. Surv. Prof. Paper 440-K, K-1–K-29.

(70) Ross, C. S. and Smith, R. L. 1955. Water and Other Volatiles in Volcanic Glasses. *Am. Mineral.* 40: 1071–1089; Day, A. L. and Shepherd, E. S. 1913. Water and Volcanic Activity. *Bull. Geol. Soc. Am.* 24: 573–606.

(71) Williams. H. 1954. Problems and Progress in Volcanology. *Geol. Soc. London, Quart. Jour.* 109; 311–332; MacDonald, G. A. 1961. Volcanology. *Science,* 133, 3454: 674–675.

(72) Morey, G. W. 1922. The Development of Pressure in Magmas as a Result of Crystallization. *Jour. Washington Acad. Sci.* 12, 9: 219–230; Verhoogen, J. 1951. Mechanics of Ash Formation. *Am. Jour. Sci.* 279, 10: 729–739.

(73) Shepherd, E. S. and Merwin, H. E. 1927. Gases of the Mt. Pelée Lavas of 1902. *Jour. Geol.* 35, 2: 110.

(74) Fries, C., Jr. 1953. Volumes and Weights of Pyroclastic Material, Lava, and Water Erupted by Parícutin Volcano, Michoacán, Mexico. *Am. Geophys. Union Trans.* 35, 4: 603–616.

(75) Sapper, K. 1928. Die Tätigsten Vulkangebiete der Gegenwart. *Zeitschr. Vulk.* 11, 3: 181–187.

(76) Ross, C. S. 1955. Provenience of Pyroclastic Materials. *Bull. Geol. Soc. Am.* 66: 427–434.

(77) Mackin, J. H. 1960. Structural Significance of Tertiary Volcanic Rocks in southwestern Utah. *Am. Jour. Sci.* 258, 2: 83.

(78) Larsen, E. S., Jr., and Cross W. 1956. Geology and Petrology of the San Juan Region, southeastern Colorado. *U.S. Geol. Surv. Prof. Paper 258*, 94 pp.

(79) Kingsley, L. 1931. Cauldron Subsidence of the Ossipee Mountains. *Am. Jour. Sci.* 5th Ser. 22, 128: 139–168.

(80) Anderson, E. M. 1936. The Dynamics of the Formation of Cone Sheets, Ring Dikes, and Cauldron Subsidences. *Roy. Soc. Edinburgh Trans.* part 2, 56: 128–157.

(81) Richey, J. E. and Thomas, H. H. 1930. The Geology of Ardnamurchan, Northwest Mull and Coll. *Geol. Surv. of Scotland Memoir*, 393 pp; Richey, J. E. 1961. Scotland: The Tertiary Volcanic Districts. *Dept. Sci. Ind. Res.*, Edinburgh, 3rd ed., 120 pp.

(82) Billings, M. P. 1945. Mechanics of Igneous Intrusion in New Hampshire. *Am. Jour. Sci.* (Daly Volume), 243A: 40–48; Oftedahl, C., 1953. Studies on the Igneous Rock Complex of the Oslo Region 13. The Cauldrons. *Norske Vidensk. -Akad. Oslo, Mat. -Nat. Kl., Skr.* 3, 108 pp.; Jacobson, R.R.E., MacLeod, W. N. and Black, R. 1958. Ring-complexes in the Younger Granite Province of Northern Nigeria. *Geol. Soc. London,* Memoir 1, 72 pp.

(83) Paige, S. 1913. The Bearing of Progressive Increase of Viscosity during Intrusion on the Form of Laccoliths. *Jour. Geol.* 21, 6: 541–549; MacCarthy, G. R. 1925. Some Facts and Theories concerning Laccoliths. *Jour. Geol.* 33: 1–18.

PLATE 1A

Parícutin, Michoacán volcanic province, 21.5 kilometers northwest of Uruapan, Mexico. On February 20, 1943, Demetrio Toral, a laborer from the village of Parícutin, was plowing land on the farm of Cuiyusuru. At 4:30 p.m. he had just completed a furrow and was about to turn his plow when the first outbreak of the volcano occurred in almost the exact furrow Toral had just drawn. The new cone reached a height of 167 meters in six days of activity showering enormous quantities of basaltic ash over the countryside. In the 1943–1952 period, the amount of pyroclastics evolved was 2.23 billion tons (metric) and the amount of lava 1.33 billion tons (metric). In these nine years of activity, aphanitic, blocky to aa andesite basalt flows with sparse olivine and hypersthene phenocrysts extended some 4 kilometers from the base of the cone shown. The flows totally covered two towns except for the church tower at San Juan Parangaricutiro (right of cone base). A slight progression toward a more siliceous lava is suggested by the rock chemistry over the period of the eruption. The main phases of Parícutin activity may be divided into three periods: (1) Quitzocho period, during which the activity was centered about the original Cuiyusuru vents and during which the volcano built its cone; (2) Sapichu period with the principal activity taking place in the later Sapichu vents and the adventitious cone; and Sapichu and (3) Taqui period, when activity was largely connected with the Taqui and Ahúan vents. See Foshag and Jenaro Gonzalez (1956). Also, a comprehensive account of the petrology and volcanology of the Parícutin volcano is given in Bulletin 965 of the U.S. Geol. Surv. in 1965. Photo: T. Nichols.

PLATE 1B

Early morning time exposure (February, 1944) recording the paths of incandescent "bomb" ejected through the air and down the sides of the cone of Parícutin Volcano, Mexico. Gray ash covers the lava hills at the base of the cone. Caption and Photo: T. Nichols.

PLATE 2A

Mount Lassen, in northern California, in eruption on August 6, 1915. This is one of the few volcanoes in the continental U.S. (excluding Alaska) to be active in historic times. Lassen Peak, seen here, is one of a series of dacite domes associated with Brokeoff Mountain, an older andesitic cone to the southwest. Lassen began to erupt clouds of steam and ash (no lava) in May of 1914, continued with many eruptions that year, and resumed its activity in May, 1915. The most intense events occurred during May 19–22, 1915, in which ash and mudflows accumulated along Lassen's slopes. Major activity persisted until June, 1917, after which the Peak entered a period of quiescence existing up to the present time. The emission of steam (possibly fed by melt waters from capping snows), and ash produced and enlarged several explosion craters near the summit. Lassen Peak is composed of hyalodacite containing glass-charged zoned andesine, quartz, green hornblende, biotite, and hypersthene in a glassy groundmass stippled with crystallites. Basic inclusions often occur in the dacite and are characterized by labradorite and calcic andesine and reddish brown oxyhornblende largely replaced by magnetite and hematite. A comprehensive account of the eruption of Mount Lassen and its subsequent hot spring activity is given by Day and Allen (1925). Photo: Geophys Lab., Carnegie Inst. Washington, D.C.

PLATE 2B

Eruption of Mount Mayon in the Philippines during the week of April 20, 1968. Cauliflowering of dust-laden gases in upper portions of nuées ardentes flowing down slopes obscures the channeling action of the lower denser portions of the flow. Mud flow deposits are further downslope from the lava flows because the mud is washed down by rain. The ash flow deposits high on the slopes of Mayon levees at the channel borders. A detailed account of the recent eruptive activity of Mayon is in the 1968 Comvol Letter, Vol II., No. 3, of the Commission on Volcanology, University of the Philippines and in the June 10, 1968 Event Report "Mount Mayon Volcanic Eruption, Philippine Islands" of the Smithsonian Institution, Center for Short-Lived Phenomena, Washington, D.C. The last eruption of Mayon occurred in 1947. Photo: W. Melson.

PLATE 3A

Nuées ardentes (glowing avalancees or ash flows) of Mounr Pelée, December 16, 1902. Mount Pelée, situated on the northern end of Martinique, produced a sequence of ash flows, the most destructive of which occurred on May 8, 1902 when at 8:02 a.m. a nuée destroyed the nearby seacoast town of St. Pierre and all but one of its 28,000 inhabitants. The cloud shown here is one of many produced within a period of 5 minutes and which rose to 4,000 meters. The flow rate was 20 meters per second, its temperature about 800°C. The lubricating fluid for the gas-glass-dust emulsion of a nuée ardente was probably mostly carbon dioxide. The photo was taken some 40 days after a dacite spine (plate 69A) began to grow in the summit crater, la Caldeira de I'Etang Sec. By May 31, 1903 the spine reached its maximum height of about 350 meters above the crater floor and 1,617 meters above sea level. A later series of eruptions in 1929–1932 produced volcanics petrographically similar to those of 1904–1905. The dome lavas are hypersthene andesites and the flow products and spines are dacitic. Rocks containing quartz appeared only in the later phases of the eruptive cycle. The most comprehensive account of the 1902 eruption is by Lacroix (1904) and that of the 1929 eruption by Perret (1937). Photo: A. Lacroix.

PLATE 3B

Explosive eruptions accompanied by well defined base surges. Left: September 30, 1965 eruption of Taal Volcano, Philippines. Right: Early October, 1957 eruption of Capelinhos Volcano, Azores. The white bar in each photo is approximately 500 meters long. A base surge is a ring-shaped basal cloud that sweeps outward as a density flow from the base of a vertical explosion column, at initial velocities usually greater than 50 meters per second. The expanding gases, which first vent vertically, with continued expansion rush over the crater lip, tear ejecta from it, and feed the gas-charged density flow. Base surges are a common feature of shallow, submarine and phreatic volcanic eruptions. The steam cloud can transport ash, lapilli, and blocks well beyond the limits of throw-out trajectories. Closer to the eruption center, a base surge can erode radial channels and deposit material with dune-type bedding (Plate 103). Caption: J.G. Moore. Left Photo: L.E. Andrews; Right Photo: U.S. Air Force.

PLATE 4A

Anak Krakatau II in the Sanda Strait between Sumatra and Java, eruption of February 9, 1929. View to southwest. Jets of pyroclastics, ash, lapilli, and blocks, 400 meters high, are well shown. Immediately prior to this eruptive phase, 11,791 explosions were recorded in 24 hours (February 3–4). By July 3rd of 1929 the island completely disappeared. In December 1929, the eruption point of Anak Krakatau shifted 600 meters to the southwest, and another 125 meters to the south on June 2, 1930. At the present time Anak Krakatau IV is over 132 meters high. The oldest Krakatau rocks consist of tridymite andesite; the eruption products of Anak Krakatau IV initially basaltic are, since 1935, made up of augite hypersthene andesites, sometimes containing olivine. A detailed account of the cataclysmic eruption of Krakatau (west of Java) in 1883 is given by Symons (1888). Photo: Indonesian Geol. Surv., contributed by R. Decker.

PLATE 4B

Pyroclastic cone in central vent of Anak Krakatau IV. Steam and ash eruption of January 12, 1960. Rakata, a remnant island of the original stratocone formed after the 1883 explosion of Krakatau, is in the left background. Rakata, with a maximum diameter of 5½ kilometers, is 813 meters above sea level. Over 100 meters of pumice occur on Rakata Island in contrast to the basaltic ash of present-day Anak Krakatau IV. Photo: R. Decker.

PLATE 5A

Surtsey, southwest of the Vestmann Islands, Iceland. The visible eruption started on November 14, 1963 and the island was born the following night. The depth to the sea floor prior to the occurrence of Surtsey was 130 meters and probably about a week was required for the volcano to reach sea level. At the time the photograph was taken (November 30), two weeks after Surtsey appeared, the island was 100 meters high and 800 meters wide. The vent was open to the sea to the southwest as is evidenced by the large white cloud of steam on the left. The darker cloud consists of basaltic tephra. On this date a rare lull of 15 minutes was recorded in the generally continuous eruptive activity. By April, 1963, the volcano built a barrier between the sea and the main vent and the permanence of the island was assured. A total of some 350 million cubic meters of lava and about 600 million cubic meters of tephra were evolved in the creation of Surtsey. An excellent photographic record of the birth of Surtsey is given by Thorarinsson (1964). Caption and Photo: S. Thorarinsson.

PLATE 5B

Surtsey, off the south coast of Iceland, October 2, 1966, illustrating the evolution of a small basaltic shield volcano following a submarine eruption (see Plate 5A). Northeast is toward the left. Northern halves of circular tephra rims (Surtur I, lower; Surtur II, upper) are vestiges of the pyroclastic stage of eruption, November 1963 through the winter of 1964. Circular central vent of Surtur II and its contiguous lava shield to the south were formed in 1964 and 1965 during a major series of effusive eruptions. Pressure ridges, tumuli, and circular collapse features visible in the southern half of the island date from this period. An effusive fissure eruption from the floor of Surtur I, yielding dark-colored lava on the top side of the island, was in progress at the time the photograph was taken. A wave-eroded remnant of Jólnir satellite volcano appears in the lower right corner. Jólnir's last visible explosive pulse occurred on August 10, 1966, nine days before the outbreak along the Surtur I fissure. Caption: J.D. Friedman, U.S. Geol. Surv.; Photo: Aerial photograph by Icelandic Survey Department, Reykjavik.

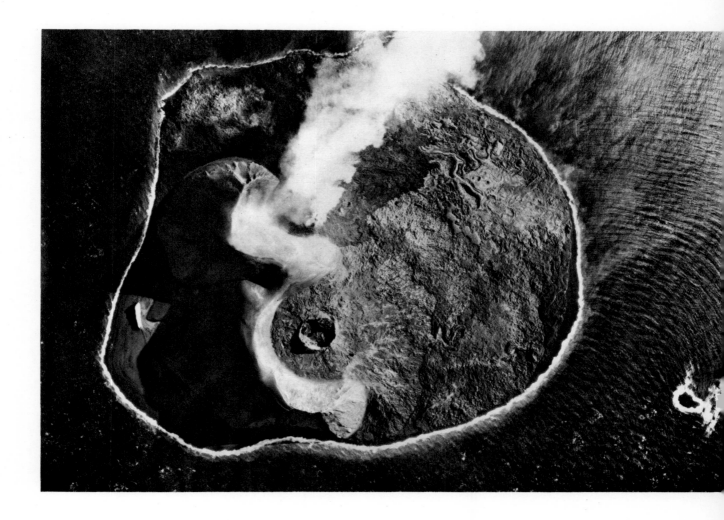

PLATE 6A

Basaltic volcano, Bayonnaise Rocks area, northern part of Izu-Mariana Arc. The island results from volcanic activity on a submarine rift. The Bayonnaise rocks may be the western rim of a submarine caldera 10 kilometers in diameter whose northeastern rim was lost by either collapse or down-faulting. To the east of the Bayonnaise Rocks, at Lat. 31° 57′ N., Long. 140° 01′ E., a new volcano shown here appeared in early February, 1946 and persisted until October, being reduced to a slight reef by December of that year. In middle September 1952 the lava dome, Myozin-syo, gradually rose at about the same point as the 1946 eruption. A part of the dome protruded as a steep spine which subsequently collapsed and submerged. The island was subsequently built up three times until final collapse in 1953. On September 24, 1952, a research boat, the Daigo-Kaiyo-maru, approaching the site was completely destroyed by a submarine eruption killing 31 research scientists and crewmen. Subsequent submarine eruptions have occurred in this area since 1953. Although hypersthene-augite dacite pumice was produced from these eruptions, the Bayonnaise Rocks themselves are made up of tholeiitic hypersthene-augite basalt. Photo: U.S. Navy, contributed by H. Foster.

PLATE 6B

Lava entering the ocean from the Island of Hawaii, 1960. Dense clouds of steam-charged ash are created as the molten lava contacts the seawater. Much beach sand is created by this process. Photo: R.T. Haugen, Nat. Park Serv., U.S. Dept. Interior.

PLATE 7A

Ash cloud of the March 30, 1956 Katmaian eruption of Bezymianny Volcano, Kluchevskaia group, central Kamchatka, Siberia. As a result of the explosion of 1956, the southeastern slope of the volcano was destroyed and the overall height reduced by 200 meters, the present height being 1,300–1,500 meters. A new dome was formed on the volcano interior. The eruptive products include andesites and dacites. The height of the cloud reached some 35 kilometers, the cloud base remaining 6–8 kilometers above the ground. Photo: I. Erova, contributed by G. S. Gorshkov.

PLATE 7B

Volcanic cloud of June 11, 1968 eruption of Isla Fernandina, Galapagos Islands. Photograph taken from Academy Bay in Santa Cruz, 140 kilometers distant. The cloud was backlighted by the late afternoon sun and was spreading at an estimated lateral speed of 80 kilometers per hour. At the time of this photograph, the diameter of the cloud was 175 kilometers and its altitude had reached between 20 and 25 kilometers. Caption: T. Simkin. Photo: J. Harte.

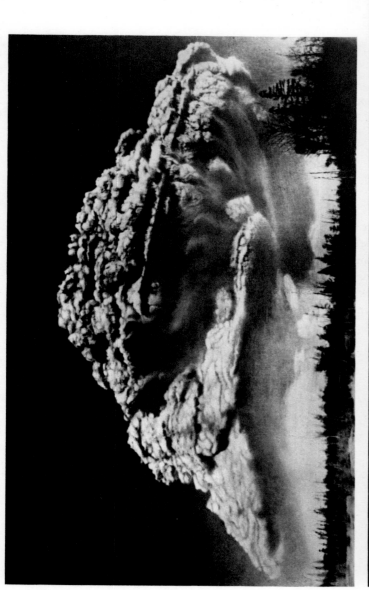

PLATE 8A

A curtain of tephra (ash) falling from eruption clouds drifting from the Island of Surtsey, November 26, 1963. Caption and Photo: S. Thorarinsson.

PLATE 8B

A whirlwind formed from a steam and ash (tephra) cloud from the nearby Surtsey volcano, off the south coast of Iceland. The funnel shown here extends to the ocean surface, so that it acts as a waterspout. The wind in this tornadic vortex can rotate in either direction, i.e., cyclonic or anticyclonic. Often, several vortices develop simultaneously from the same cloud. Whirlwinds can be produced in eruption clouds when special conditions, including a concentration of thermal and kinetic energies from updrafts of hot gases and sufficient angular momentum in the moving air mass, are obtained within the clouds. Photo: Courtesy U.S. Navy, contributed by J. Hughes.

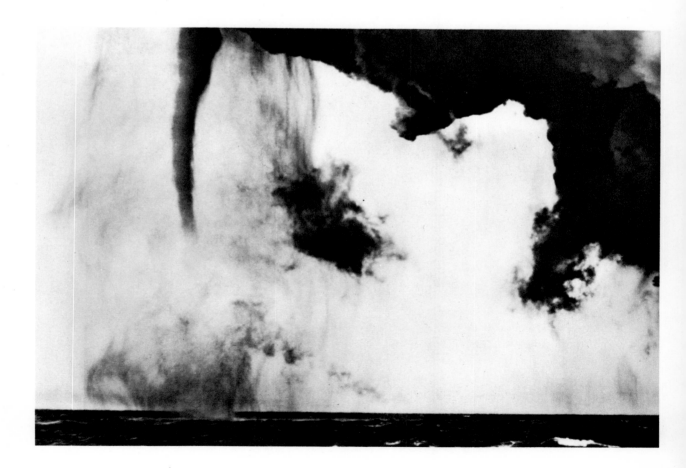

PLATE 9A

Lightning associated with volcanic eruption at Surtsey (near center of horizon) in the Vestmann Islands, southwest of Vestmannaeyjar on the southern coast of Iceland. Almost all volcanic eruptions on earth, especially those involving dust, are accompanied by electrostatic effects. Electrostatic measurements were made from February through July of 1964 of the positive charge in the volcanic ash clouds of Surtsey. The volt per centimeter values ranged from one to 300 with a fine weather potential gradient at the ocean surface of 1.3 volt per centimeter (Anderson *et al.*, 1965). The concentration of charge which occurred in Surtsey ranged from 10^5 to 10^6 elementary charges per cubic centimeter at the time of eruptive bursts from the volcano orifice. The cause of the charge separation mechanism is believed by Anderson *et al.* to be due either to contact of seawater with magma or lava, or collision of particles carried in the erupting gases with each other on the walls of the vent. Many land-based volcanoes such as Irazu or Parícutin are not in contact with seawater and exhibit positively charged clouds similar to those at Surtsey. Photo: S. Jonasson.

PLATE 9B

The phenomenon of "flashing arcs", as they would appear if caught by camera, during strong explosive eruptions of Vesuvius on April 7, 1906. This rare feature is produced by compression of air around local explosion centers at the top of a volcano vent. As the compression wave moves spherically outward at the speed of sound, a pressure discontinuity occurs at the wave front such that the air behind the interface is at a higher density than the, as yet, undisturbed air immediately ahead of the advancing wave. The density contrast causes local differences in light refraction that lead to illumination of the interface. Perret (1912) observed this phenomena at the April 7th eruption shown, as well as at later eruptions at Vesuvius and at Etna. Relative to Plate 9B, Perret states "photography having failed, I have permitted myself . . . to trace circles with aniline upon the negative which gives a crude idea of the arcs as though caught at a given instant of their flight." Photo: Geophys. Lab., Carnegie Inst., Washington, D. C.

PLATE 10A

Lava fountaining in the summit of Mauna Loa, 1949. The eruptions which began on January 6, lasted 145 days producing 59 million cubic meters of hypersthene-bearing basaltic lava covering 14.5 square kilometers. A comprehensive account of this eruption is given by MacDonald and Orr (1950). Photo: U.S. Air Force.

PLATE 10B

Fire fountaining on rift zones on the southwest flank of Mauna Loa volcanic complex. Earlier flows can be distinguished by overlap patterns and albedo. Fire fountaining on the flanks of Hawaiian calderas are often related in time to summit eruptions. Photo: U.S. Air Force.

PLATE 11A

Chain of spatter cones with Alae Crater in background. Spatter cones resulted from February 22, 1969 eruption in Hawaii Volcanoes National Park. Cones were all but obliterated by the May 24, 1969 eruption in nearby areas. Photo: D. W. Reeser, Nat. Park Serv., U.S. Dept. Interior.

PLATE 11B

Basaltic lava issuing from East Rift Zone of Kilauea, Hawaii on May 24, 1969. Chain of Craters Road is blocked by the lava flow near Aloi Crater. The lava skin is extremely vitreous with a fragile, often spinose outer layer. Photo: D. W. Reeser, Nat. Park Serv., U.S. Dept. Interior.

PLATE 12A

Active lava lake on the floor of the summit crater Halemaumau, Kilauea Volcano, Hawaii. The thin crust continuously cracks as it spreads across the lake, revealing glowing lava below. A boiling area of fountaining lies near the center of the lake; pahoehoe flows are fed by lava cascading over the levee walls. Caption and Photo: R. S. Fiske, U.S. Geol. Surv.

PLATE 12B

The fire-pit of Halemaumau seen at night during lava lake activity in 1967 within Kilauea Volcano, Hawaii. The cloud column is composed of steam and other gases and contains essentially no ash. Caption and Photo: R.S. Fiske, U.S. Geol. Surv.

PLATE 13A

Cascade of molten basaltic lava in 1963 emanating from a fissure in the wall of Alae Crater of the East Rift Zone of Kilauea, Hawaii. The lava is forming a lake on the floor of the inner crater. The diameter of the inner crater in a northwest direction is 420 meters, that of the other crater in a north-west direction 650 meters. A fire fountain occurs to the left of the cascade. Photo: G. A. Smathers, Nat. Park Serv., U.S. Dep. Interior.

PLATE 13B

Fire fountain activity in Halemaumau, a pit crater on the floor of the Kilauea Caldera, Hawaii. Halemaumau is 975 meters in diameter and about 131 meters deep. Its depth fluctuates because of infilling and draining of lava supplied or withdrawn from a shallow magma chamber. The lava is an olivine basalt containing about 50% SiO_2. The photograph was taken at 7:15 p.m., July 12, 1912 with the camera shutter open for 2/3 seconds which produced the light streaks on the surface of the flowing molten lava streams. Fire fountains can exceed 300 meters in height. Photo: Geophys. Lab., Carnegie Inst., Washington, D.C.

PLATE 14A

Pahoehoe flow forming from overflow of active lava during eruption of 1967 in Halemaumau Crater, Kilauea Volcano, Hawaii. Caption and Photo: R. S. Fiske, U.S. Geol. Surv.

PLATE 14B

General view of a lava lake, enclosed in a lava ring, on the floor of the Halemaumau pit crater in the Kilauea Caldera, Island of Hawaii, May 28, 1912. The lake is rimmed by a spatter wall, up to 2 meters high (inset, upper left). Note the lobes of overflowed lava, the several rim breaches from which lava channels emerge, and the active lava fountain that has developed in the thin crust of the lake. Photo: Geophys. Lab., Carnegie Inst. Washington, D.C.

PLATE 15A

Aerial photograph of Hekla, Iceland, August 13, 1960, showing (left to right) southwest flank eruption fissure and lava flows of 1947–48; Shoulder Crater (Axlargígur); section of northeast-trending eruptive fissure Heklugjá; and Summit Crater (Toppgígur). White areas are perennial firn fields and glaciers; large feature at top center is Litla Hekla (from Friedman *et al*, 1969). Caption: J. D. Friedman, U.S. Geol. Surv.; Photo: U.S. Air Force.

PLATE 15B

Aerial infrared image (post-sunset) of the same part of Hekla as above (A), August 29, 1966; 4-5.5μ wavelength region. Light areas represent high thermal emission. The following thermal anomalies are visible: (1) anomalies due to vapor emissions and fumarolic activity along northeast-trending eruption fissure Heklugjá, and (2) curvilinear fractures and fumaroles in Shoulder Crater and Summit Crater. Dark areas are perennial firn fields and glaciers. Cool cloud (dark) obscures extensions of eruption fissure on southwest shoulder of volcano (from Friedman, *et al*, 1969). Caption: J. D. Friedman, U.S. Geol. Surv.; Photo: Infrared image: R. S. Williams, Jr., Air Force Cambridge Research Laboratories.

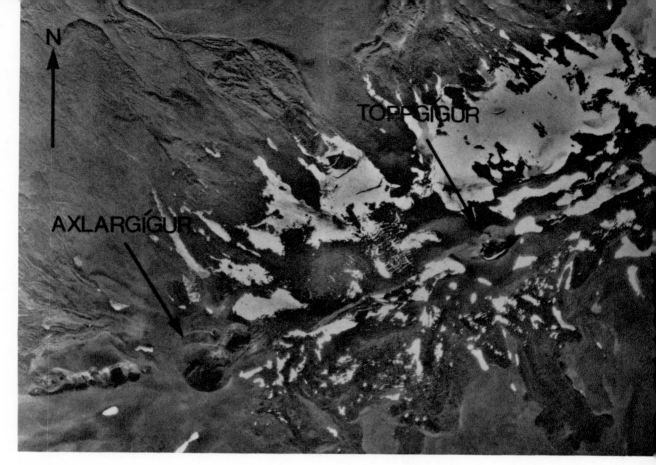

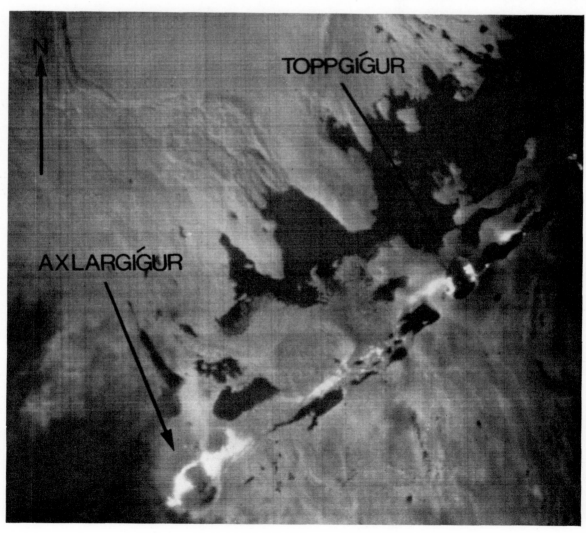

PLATE 16A

Oblique view to the north of much of the north-central part of New Mexico into southern Colorado, taken from the Apollo 9 spacecraft at an altitude of 170 kilometers. The Jemez mountains (northwest of Santa Fe) which contain the Valles Caldera (Plate 16B) appear near the center of the photo. The small, dark, roundish areas to the upper right are a group of Tertiary volcanoes west of Taos and the Rio Grande, some of which have a subdued shield volcano form. The Mount Taylor Peaks volcanic field is seen to the southwest (lower left) of the Jemez Mountains. Much of the upland areas (including the Sangre de Cristos Mountains on the right) were snow-covered during this March, 1969 flight. The darker areas in the lower center are intermontane basins. Photo: NASA.

PLATE 16B

Oblique view of a relief map of the Jemez Mountains, New Mexico. The Valles Caldera (center), the type resurgent calderon (Smith and Bailey, 1968), has a central structural dome produced by post-collapse uplift of the caldera floor. Peripheral to the central structural dome is a ring of younger rhyolite volcanic domes. The dissected plateaus east, west and north (right) of the caldera are underlain mainly by ash flows (Bandelier Tuff; see Plate 167), eruption of which caused collapse of the caldera. The diameter of the caldera is about 22 kilometers. Caption and Photo: R. Bailey, U.S. Geol. Surv.

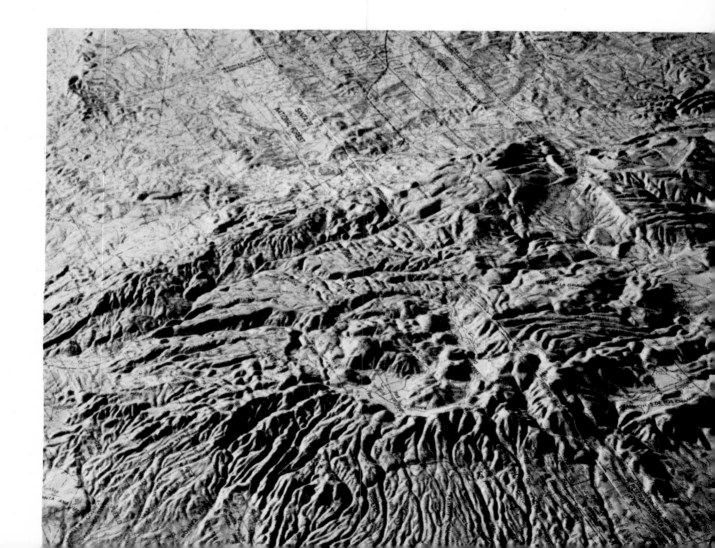

PLATE 17A

Model of Ambrym Island, central New Hebredes Islands. The Island of Ambrym measures 45 kilometers from east to west and 30.5 kilometers from north to south. It is built up of mostly labradorite basalts containing augite. Some of the older basalts contain olivine; others are very rich in labradorite, making the rock appear less mafic than it really is. The Island contains a caldera measuring 9 by 12 kilometers with walls 40 to 60 meters high; the floor is about 650 meters above sea level. Within the caldera are two large secondary cones with well-formed summit craters: Mount Marum the higher, and Mount Benbow, the more active of the two. Several aligned vents may be seen on the peak to the left. Photo: supplied by G. J. H.; McCall model prepared by P. J. Stephenson.

PLATE 17B

In this aerial photograph of Ambrym, the structural control of the location of Mount Benbow (bottom) and Mount Marum (top) is shown to be on a fissure zone which is also deforming the caldera wall and outer slope. The internal craters are elongated in the direction of this trend, a pattern that must be primary and not due to secondary distortion. In Marum, the present activity stems from a small subsidiary cone within it on the east (upper) rim. Step faulting in the caldera wall indicates Ambrym to be primarily a subsidence caldera. Photo: Royal Air Force, contributed by G. J. H. McCall and P. J. Stephenson.

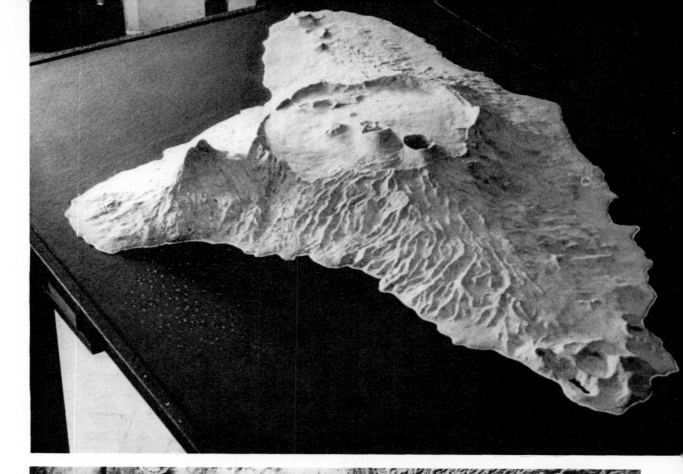

PLATE 18A

Isla Fernandina, Galapagos Islands, Ecuador. This uncontrolled mosaic shows a summit caldera with an off-center breached cone on its floor. The caldera is 7 kilometers long. A violent explosion and collapse in 1968 have since altered the interior of this caldera (Plate 19A). Note the numerous lava flows along the flanks of this shield volcano. Many of these flows, of various ages, emanate from fissures near, and subparallel to, the caldera rim. Individual and rows of adventive cones also have developed on the volcano flanks. Photo: U.S. Air Force in cooperation with the government of Ecuador contributed by J. Green.

PLATE 18B

Air photo mosaic of Fernandina Caldera, Galapagos Islands, in 1947. Note the concentric rows of spatter cones around caldera rim. The caldera floor, here roughly 600 meters below the rim, was covered by fresh lava in 1958 and subsided another 300 meters in 1968. The 1968 collapse took place over a two week period beginning with the eruptive cloud shown on Plate 7B (see also Plate 19). Caption: T. Simkin; Photo: U.S. Air Force.

APPROXIMATE SCALE (KM)

N

PLATE 19A

Oblique air photography of Fernandina Caldera, Galapagos Islands, taken three weeks after the start of the 1968 collapse, viewed from the southeast. Prominent features shown in Plate 18B, such as the northwest and southeast benches and the central tuff cone, can also be seen in this photograph despite approximately 300 meters subsidence of the southeast caldera floor. Note the dust from rock avalanches on the oversteepened walls. Low clouds obscured the volcano's flanks when the photo was taken. Caption: T. Simkin; Photo: Sgt. G. Pierson, U.S. Air Force.

PLATE 19B

Panorama of Fernandina Caldera before and after 1968 collapse. Scales of both photographs are approximately equal, as shown by the 300-meter interval between the rim and northwest bench (right). Note the tilting and subsidence of the caldera floor in the post-1968 photo. That the central cone remained intact is remarkable. Both panoramas are viewed from the northeast rim of the caldera, but the lower photo of July 12, 1968 was taken from a position slightly northwest of the upper photo, taken August, 1966. Caption: T. Simkin; Upper Photo: P. Colinvaux; Lower Photo: T. Simkin.

PLATE 20A

Vertical photograph taken with a 70 millimeter Hasselblad 500C camera (80 millimeters focal length lens) from about 225 kilometers altitude during the Apollo 9 earth orbiting space flight. The area shown is in the western Tibesti massif in the Republic of Chad. This massif, like several others in North Africa, is essentially a large Precambrian block epeirogenically uplifted and capped with Cenozoic volcanics. The volcanics shown here are chiefly ignimbrites (light-colored units); the black flows at bottom (Pic Toussidé (A) are andesite. The crater with the white floor just above Pic Toussidé is the Trou au Natron (B), a caldera 6 kilometers wide and about 950 meters deep, with a sodium carbonate playa on its floor. The other circular structures are also Cenozoic calderas of even greater size: Tarso Abeki (C), Tarso Voon (D), Tarso Toon (E), and Tarso Yega (F). Tarso Yega is the largest, with a 22 kilometer diameter. Caption: P. Lowman; Photo: NASA.

PLATE 20B

Views of the Trou au Natron and the Ehi Toussidé in the western Tibesti, Chad, in northern Africa. The most celebrated volcano of the Tibesti, the Trou au Natron is an explosion-and-collapse structure 6 to 8 kilometers in diameter and up to 1 kilometer in depth. Sodium carbonate (natronite) lines its floor, on which four smaller andesitic volcanoes are located. The Pic Toussidé, at 3,265 meters the second highest peak in the Tibesti, appears in the background. This is a trachyandesitic volcano with pyroclastic material on top (lighter patch in the photograph). Caption and Photo: A. Pesce.

PLATE 21A

Partial aerial view of the ignimbritic caldera of Tarso Voon, some 18 kilometers in diameter, in the central Tibesti, Chad. The floor at the center of this caldera has subsided at least 1,000 meters. The ignimbrite beds along the rim show marked inclination towards the middle of the caldera. The thermal springs of Soborcm, the only major trace of present volcanic activity in the Tibesti itself, are located about 5 kilometers to the west of Tarso Voon. Caption and Photo: A. Pesce.

PLATE 21B

Caldera of Oldoinyo Lengai (The Mountain of God), Rift Valley, south of Lake Natron, Tanzania. The caldera is 2,880 meters above sea level, 1,942 meters above the surrounding plain, and 2,278 meters above Lake Natron, 16 kilometers to the north. The maximum outer slope angle is 42 degrees. Recent ash cones are shown on the floor. One of its past features was a 36-meter high needle composed of soda carbonate located on the northern edge of the summit; it was destroyed in the 1913–1915 eruptions. The cone is a composite of ijolitic and nephelinitic pyroclastics containing ejectamenta of urtite, ijolite, melteigite, jacupirangite, biotite pyroxenite, wollastonitite, sovite, fenite, and various lavas.

The flow sequence from youngest to oldest includes:

1. Yellow ijolitic tuffs, agglomerates, and interbedded lavas.
2. Gray tuff and agglomerates of the adventive cones, craters, and tuff rings.
3. Black nephelinite tuffs and agglomerates.
4. Melanephelinite flows.
5. Variegated tuffs of the active crater, soda deposits of the summit area, black and gray ash.
6. Very recent lavas of the active crater.

The interlayered flows change with decreasing age from phonolite + nephelinite to nephelinite to melanephelinite, displaying a distinct decrease in SiO_2 with time. Eruptions were recorded in 1880, 1894(?), 1904, 1915, 1917, 1921, 1926, 1940, 1941, 1954–1955 and 1960, when soda-rich and essentially silica-free lava was extruded from the vent of the northern crater both as pahoehoe and aa. Soda-carbonate ash had been noted in earlier eruptions. The 1917 eruption scattered ash and "soda" as far as 48 kilometers from the volcano. Petrographically the lava consists of 1 to 2 millimeter phenocrysts of a new Ca-Na-K carbonate mineral set in microcrystalline matrix of thermonatrite (hydrated sodium carbonate), fluorite, and pyrite (?) (Heinrichs, 1966, p. 490–491). Photo: C. M. Bristow, supplied by G. J. H. McCall.

PLATE 22A

Vertical aerial photos (1:50,000) of the Quaternary Fantale Volcano Complex, lying along the main Ethiopian rift zone of the East African rift system, located in western Ethiopia. The caldera is approximately 3-½ by 4-½ kilometers and is elongated in the direction of the fissure zone. Silicic to intermediate lavas are emitted by this volcano. Note the lava channels and levees along the north (right) slopes and the recent flow field to the northeast (lower right) marked with curved flow ridges. Caption: I. Gibson; Photo: Courtesy Imperial Highway Authority of Ethiopia.

PLATE 22B

The Tengger Caldera in East Java, view to the north. The caldera is about 16 kilometers in diameter and 0.5 kilometers deep. Seven eruptive centers appear on the floor of the caldera which are localized on east-west and NNE-SSW fracture zones, the latter localizing Bromo, the active steaming volcano in this view. Bromo is 2,329 meters above sea level and 200 meters above the caldera floor. Although Bromo has not produced lava flows in historic time, large amounts of volcanic bombs, ash and lapilli have been produced; the products are mostly olivine-hypersthene basalt with augite and labradorite-bytownite plagioclase. The Bromo crater, with a north-south diameter of 800 meters and an east-west one of 600 meters, contains a terrace indicating a migration of its eruptive center to the north. To the right of Bromo is the volcano Batok (Plate 179B). The caldera floor contains a large ash-filled crescentic area called the "Sandsea". Not seen below the foreground is the stratovolcano Semeru on the same zone of structural weakness that passes through the Tengger Caldera. Photo: Dept of Information, Republic of Indonesia.

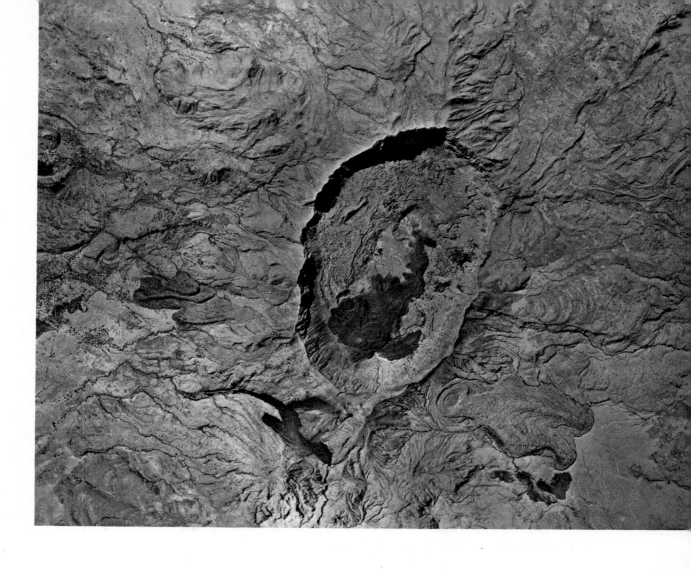

PLATE 23A

Vertical airphoto of O Shima Volcano on Island of O Shima in Tokyo Prefecture, Japan taken in 1947. The main crater walls form a somma which encloses a caldera about 3 kilometers in diameter open to the northeast. The central cone of Mihara Yama, in the south center of the caldera, is 800 meters wide and rises to an altitude of 755 meters; the pit crater within this cone is 300 meters wide. This crater, which was floored with dark lava from an eruption in 1941, underwent renewed eruptions in 1950, 1951, and 1953. In this period basaltic lava was erupted both passively and explosively. Caption: H. Foster, U.S. Geol. Surv. Source of photo unknown.

PLATE 23B

Oblique aerial view of Mihara Yama central cone within O Shima Caldera showing activity on March 23, 1951. The Kagamibata basalt flow is spilling over the crater crest and advancing northward; steam is shown venting from a 1951 cinder cone on the south end of the crater floor. Inner pit obscured. Caption: H. Foster, U.S. Geol. Surv. Photo: Asahi Shimbun.

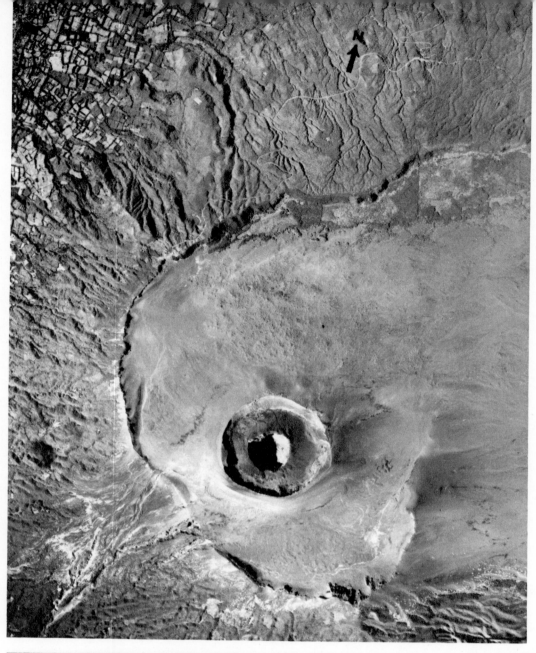

PLATE 24A

Crater Lake, Klamath County, Oregon. This aerial photo (1:54,000) shows the lake, which is 10 kilometers in diameter, and part of the eastern and western slopes of ancient Mount Mazama, the composite volcanic complex that existed at this site prior to the formation of the present caldera. The U-shaped valleys on the slopes of the volcano are now cut off by the walls of Crater Lake. These valleys were carved by alpine glaciers that mantled the upper slopes of Mazama at the time of the eruptions that accompanied the subsidence of this volcanic complex to produce the caldera now occupied by Crater Lake. Wizard Island (right end of lake), a small volcanic cone, evolved long after the caldera formed. Caption (modified) and Photo: From U.S. Geol. Surv. Prof. Paper 590.

PLATE 24B

Northwest inner wall of Crater Lake, Oregon. Crater Lake resulted from the explosive eruptions and collapse of the volcanic complex Mount Mazama. The concept of a single explosion that obliterated the summit of a pre-existing volcano is no longer entertained. Multiphase eruptions and collapses over a long time period account for the present topography of Crater Lake which incidentally is the deepest lake in the United States (580 meters). On the rim on the far left is Hillman Peak, the highest point on the rim, 610 meters above lake level. It is the remnant of a parasitic cone composed of andesite flows and pyroclastics. Llao Rock appears on the far right; it is composed of vesicular vitric dacite. The Devil's Backbone, a vertical dike of dacite, appears in between Hillman Peak and Llao Rock on the caldera wall. Wizard Island in the right center is a cinder cone of basalt and basaltic andesite that rises 230 meters above the lake. Note the blocky lava flow that extends toward shore from this post-caldera cone. Photo: G. A. Grant, Nat. Park Serv., U.S. Dept. Interior.

PLATE 25A

Nested calderas of the Kuriles, Siberia. The photo shows the inner caldera of the Zavaritski Caldera on the south half of Simushir Island. The overall diameter of the outer caldera is 6 kilometers and contains several craters each inside the other. The outer caldera is open to the Okhotsk sea, and its rim is up to 600 meters above sea level. The inner crater is not dissected and is about 2.5 by 3.5 kilometers wide with a rim height of up to 625 meters above sea level. The inner walls of the inner caldera are extremely steep. A small breached scoria cone is located; at the north bank of the inner caldera, the basement rocks of the Zavaritski Caldera are Tertiary volcanics; grandiorite xenoliths are also found in the pyroclastic products. The caldera itself is on a major volcano-tectonic fault zone. The eruption of 1957 produced a dome within the inner caldera of the Santorin type. The lavas are essentially basaltic although andesites and ignimbrites are present. The slopes of the flanks of the inner caldera are covered with a mantle of slag bombs and lapilli. Photo: G. S. Gorshkov.

PLATE 25B

Aerial view northwestward over the northwest part of the Island Park Caldera, southestern Idaho. The caldera is about 30 kilometers in diameter and is defined along its western half by a semicircular scarp about 200 meters high. The rim is formed of welded rhyolite tuff; the hills on the left are extrusive rhyolite domes. The park is floored by basalt flows, and other basalt flows from outside the caldera lap onto the outer rim. Caption and Photo: W. Hamilton, U.S. Geol. Surv.

PLATE 26A

An aerial oblique view from the northeast showing the broadly convex upward profile of the basaltic shield volcanic complex of Mauna Loa, Island of Hawaii. The length of the profile at the horizon is approximately 20 kilometers. The summit caldera of Mokuaweoweo appears to the right of the steam coming from one of the pit craters along the southwest rift zone. The main fractures of the northeast rift zone, from which numerous fissure eruptions have flowed, cross the photo from upper right to lower left starting at the north (right) end of Mokuaweoweo. The upper slopes of Mauna Loa are snow-covered in this view. Photo: U.S. Air Force.

PLATE 26B

Mokuaweoweo Caldera at the summit of Mauna Loa volcanic complex, Island of Hawaii. The caldera measures 4.8 by 2.4 kilometers. The boundary cliff on the western side is 180 meters high. Note the beheaded lava flows in the right foreground. Caption and Photo: J. G. Moore, U.S. Geol. Surv.

PLATE 27A

Kilauea Caldera, Hawaii. Oblique view to north. The caldera, 4 kilometers in maximum diameter, is bordered by fault scarps and floored by lava flows erupted within the last century. Along the western and northern edges are step faults. Near the southwest edge is the "Fire Pit" (Halemaumau) about 152 meters deep; it is the focus of Kilauea's eruptive activity. The photograph taken in late 1954 shows the spatter cone of the 1952 eruption which stands above the floor of Halemaumau. The floor of Halemaumau, covered with lavas formed in 1952 and 1954, has been considerably modified since this photograph was taken. A fissure that extends east-northeast from the northeast side of the crater fed the 1954 lava flow whose dark color on the aerial photograph contrasts with the surrounding lighter-colored weathered flows. Open fissures that extend southwest from Halemaumau are the surface expression of the north end of the southwest rift of Kilauea. Mauna Kea is in the background. The geology of this area is described by Macdonald and Eaton (1957). Caption modified from Denney, et al. (1968, pp. 33–34). Photo: U. S. Geol. Surv.

PLATE 27B

Haleakala, the easternmost of two volcanic structures in Maui. Haleakala is crossed by an east by northeast rift zone which defines the alignment of cinder and spatter cones and spatter ramparts on its floor and flanks. Most of the vents are of the Hana volcanic series and many are of the older Kula volcanic series. The rift cuts across the entire caldera of Haleakala shown here, which is 9.5 kilometers long, 3.5 kilometers wide and 600 meters deep. The caldera has been modified by headward erosion of two large valleys, the floors of which have been buried by accumulations of Hana volcanic rocks. Undoubtedly some explosivity and much subsidence have also shaped this caldera. A row of andesitic scoria cones on the right are among several series of adventive cones developed on the caldera flanks. Flows of different age can be distinguished in the photograph. The most abundant rock of the oldest Honomanu series is olivine basalt; that of the overlying Kula series is andesine basalt and its variations, and that of the youngest Hana series is olivine basalt. Only one volcanic eruption in Haleakala Caldera has occured in historic time, the lava flow of about 1750. Photo: J. W. Larson, Nat. Park Serv., U.S. Dept. Interior.

PLATE 28A

Shishaldin volcano, 2,858 meters above sea level, Unimak Island, Alaska. The cone is one of the most symmetrical in the entire Aleutian chain. Over 25 eruptions have been recorded since 1775 A.D. Photo: U.S. Navy, contributed by R. L. Smith.

PLATE 28B

Iliamna Volcano, 225 kilometers NNE of Katmai Caldera, western Alaska. The stratocone, 3,073 meters high, has had seven recorded eruptions. A steam blast is shown in the photograph. Photo: U.S. Air Force, contributed by R. Katchadoorian.

PLATE 29A

Mount St. Helens, a recent volcano in the Cascades of the state of Washington, shown here in an aerial photograph. This symmetric volcano is about 1,900 meters above the surrounding hills on the west flank of the Cascade Range. The present cone may have been built during the last 1,000 years, and the small glaciers on its steep slopes have not yet cut deep valleys. The areas of sparse vegetation on the lower slopes of the volcano have been devastated by debris flows or lahars within the past century. Photo width covers 9.6 kilometers. Caption and Photo: U.S. Geol. Surv. Prof. Paper 590.

PLATE 29B

The Barcena volcano (previously called Boquerón) of the eastern Revillagigedos Islands off the west coast of Mexico. The volcano erupted on Isla San Benedicto on August 1, 1952. The ash and cinder cone shown here grew to a height of approximately 381 meters by August 12. The diameter of the cone base is 700 meters. Vulcanian-Peléan type eruptions of pyroclastics continued into September. In November viscous andesite or trachyandesite lava partially filled the crater, and in December block lava broke through the base of Barcena and flowed into the sea until March 1953, extending the coastline 650 meters. At this time visible activity stopped except for steaming fumaroles, and these became inactive in 1954. A study of the rate of erosion of the tephra and lava of this volcano has been made by Richards (1960). Photo: U.S. Navy.

PLATE 30A

Hekla, southern Iceland, viewed from the west-southwest on August 13, 1968. (See Plate 15). A linear, polygenetic composite volcano, Hekla is built up along a typical Icelandic eruption fissure, Heklugja, which extends along its crest line. Toppgígur, the Summit Crater (highest point on the crest line) and Axlargígur, the Shoulder Crater (right background, just above the break in slope) were both active in 1947–48 when they erupted dacitic-andestitic tephra, covering slopes in foreground and middleground, followed by a lava eruption from the entire length of Heklugja. Apalhraun siliceous flow in foreground originated in the 1845 eruption. Caption and Photo: J. D. Friedman, U.S. Geol. Surv.

PLATE 30B

The Skjaldbreit lava cone or "dyngja", located 60 km east-northeast of Reykjavik, Iceland. This symmetrical structure resembles, on a smaller scale, some of the larger shield volcanoes of Hawaii and, like those, forms piling up of successive outpourings of fluid basic lavas. Skjaldbreit is nearly 10 km wide at its base, 1050 meters high, and has a typical flank gradient of about 200 meters/km. Photo: S. Thorarinsson.

PLATE 31A

El Misti volcanic complex, in the Province of Arequipa, Peru, altitude 5,825 meters. This andesitic structure, with a summit caldera enclosing an inner cinder cone, is in a state of mild fumarolic activity. The upward concavity of its slopes is characteristic of many volcanic complexes. In some, this concavity results from deposition of ash or outflow of lava along the lower slopes; at El Misti, the profile reflects erosion of the upper slopes (note dissected gullies) and deposition of the debris as fans along the base by both stream and landslide processes. Photo: Courtesy F. M. Bullard.

PLATE 31B

Merapi stratocone, central Java. This active volcano 2,911 meters high is situated at the intersection of regional faults, a transverse one which separates East Java from Central Java and a longitudinal fault through Java. An older part of the complex, Batalawang, can be distinguished from the active Merapi cone. The summit of Merapi once contained a crater (from 1872 to 1883) and at other times plug domes until December 30, 1930. During the many eruptions of Merapi, glowing clouds or nuées ardentes have destroyed the slopes and the foot of the mountain out to a distance of 12 kilometers, often with great loss of life. Mudflows often followed the nuées adding to the destruction of life and property in the area. All eruption products of the present cone are vitrophyric augite-hypersthene andesites, with or without hornblende. Photo: Dept of Information, Republic of Indonesia.

PLATE 31C

Farallon de Pajaros, lat. 20.33°N., long. 144.54 E., Northern Marianas in the Pacific. A seismic tremor lasting two days was reported on March 11, 1969 near this volcano. A previous event occurred in March 1967. Andesitic lava flows, which have been extruded from flank vents, have well-defined termini at the volcano base. Photo: U.S. Navy.

PLATE 31D

Mount Egmont (2,530 meters), New Zealand, seen from the south, showing the parasitic cone of Fantham Peak (lower right). The main crater was the last active vent. A small tholoid appears in the summit crater and a viscous coulee or overflow is just visible in the upper left. Caption and Photo: S.N. Beatus, New Zealand Geol. Surv.

PLATE 32A

Asama-yama, one of Japan's most active volcanoes, on Honshu Island. The central cone (lower left center) has a deep crater and is bordered on the west by a partial somma or caldera rim. Recent ash covers most ground near the crater. Farther away are pyroclastic deposits, loose and gullied (upper left quadrant of photo); welded deposits (northeast, upper right) are much less dissected. Ko-asama-yama, a conical hill (southeast quadrant, lower right) is a dome of viscous dacite lava. The Onioshidashi-iwa lava flow of 1783 (north from central cone) is a hypersthene-augite andesite, with a rough, blocky surface. The flow partly fills the head of a flat-floored, steep-sided trench (so-called Jambara ditch), carved by a nuée ardente that immediately preceded the lava. Photo width covers 8.5 kilometers. Caption and Photo: U.S. Geol. Surv. Prof. Paper 591.

PLATE 32B

Mount Fujiyama (Fujisan), is a composite volcano covering an area of about 50 by 35 kilometers near the southern coast of southern Honshu. The volcano, about 100 kilometers south of Tokyo, is a simple symmetrical cone of basaltic lava with a summit crater 700 meters across and 100 meters deep; the highest point is 3,776 meters above sea level. The slope of the cone is about 35 degrees near the summit and decreases toward the skirt, made up of nearly horizontal lavas and mud flows. The volcano basement is made up of Misaka lavas and Pliocene sediments of the Asigara Group. Fujiyama is actually the product of two overlapping bodies, the old Fuji volcano to the northeast and the recent onlapping Fuji volcano to the southwest. The old Fuji volcano was partially eroded prior to the evolution of new Fuji which has at least 60 side vents on the flank of the main cone. These vents are cinder cones often associated with lava flows. H. Tsuya believes that the side vents were formed at points where dike swarms radiating at relatively shallow depths intersect the surface—a feature also seen on Hawaiian volcanoes. The last recorded eruption at Fujiyama was in 1707, in which fragments of hypersthene augite-olivine gabbro were ejected. Compared with basalts of the Izu Peninsula and Highlands, the Fuji basalts are relatively high in aluminum and alkalies but low in calcium. The Fujiyama lavas, both old and new, are olivine basalt, augite-olivine basalt, and hypersthene-augite-olivine basalt. Photo: Japan Air Lines.

PLATE 33A

Aligned volcanoes of Tongariro National Park, North Island, New Zealand. North Crater, filled to the brim with basaltic lava, is in the foreground. A well-developed pit crater occurs in the lava fill. Beyond is the multiple volcano, Mount Tongariro (1,986 meters) with its symmetrical crater. Behind Tongariro is Ngauruhoe (2,291 meters) which in 1949 and 1954 emitted hot avalanches and lava; the summit crater is currently active. Ruapehu (2,797 meters) is in the distance with its permanent snow field and hot crater lake at its summit; it was last in eruption in 1945–46. All volcanoes consist of basaltic and andesitic lavas collectively called basaltic andesite. Photo: S. N. Beatus, New Zealand Geol. Surv.

PLATE 33B

White Island, Bay of Plenty, New Zealand. White Island lies on the western margin of a major northeast trending graben extending to Mount Ruapehu 240 kilometers to the southwest. The highest points on the volcano, Mount Gisborne (344 meters), is on the western margin of the crater. Active solfataras and much hydrothermally altered rock typify the crater interior. In September 1914 a great landslide of altered andesite was detached from the western wall and crossed the crater floor in the manner of a mudflow and mostly came to rest on the western portion of the floor to a depth of about 70 meters. Some of the fine debris traveled further and formed a flattish lahar deposit about 17 meters thick from which hillocks composed of large blocks of andesite project up to 14 meters high. The hillocks are most abundant in the southern side of the crater floor. The volcano is formed of andesitic tuffs, agglomerates, lava flows and dikes, the fragmental rocks predominating. The crater is surrounded on all sides by steep cliffs except at three places on the east—Crater, Wilson, and Shark Bays. Over 10,000 tons of sulfur have been shipped from this Island. A detailed account of the volcano is given by Hamilton and Baumgart (1959). Photo: S. N. Beatus, New Zealand Geol. Surv.

PLATE 34A

Aerial photo of the present-day Mount Vesuvius, southeast of Naples, Italy. The summit crater of the central cone of Vesuvius, now over 1,180 meters above sea level, is 0.8 kilometers in its longest (northwest) dimension. The rim of Monte Somma (the pre-Vesuvian stratocone, destroyed during the eruptions of 79 A.D.) lies about 1.5 kilometers to the north; the upper parts of its outer slopes are dark because of vegetation cover. Remnants of many older lava flows from Vesuvius are visible around the crater, especially to the south and west. The dark colored 1944 lava flow moved northward from the crater base into the Atrio del Cavallo, up to the inner walls of Monte Somma, and then flowed westward past the Volcano Observatory to the town of San Sebastiano (upper left). Photo: Italian Air Force. Inset shows crater in eruption (1944) seen from the air, looking to the southeast. The somma or remnant of the flanks of ancient Monte Somma appears to the left. A lobe of flowing lava, moving westward, emanates from near the base of the Vesuvius cone. Although the earliest products of Vesuvian eruptions were trachytes, the older somma is composed of phonolitic leucite tephrites, followed by eruptions of leucite tephrites composing the younger somma. In historic eruptions including the 1944 flow, all products are tephritic leucitites. Photo: U.S. Navy.

PLATE 34B

Stromboli, the northeasternmost island of the Lipari Group, north of Sicily. The town of San Vincenzo lies along the northern shore 5 kilometers from its southern tip. The central summit crater contains three active vents. To the west (bottom), lies the Sciara del Fuoco, an indentation into the slopes covered by mixed landslide and lava flows. Photo: Italian Air Force.

Inset is a view from the northwest of Stromboli, whose summit is 925 meters above sea level and more than 2,900 meters above the sea floor. The furrowed slopes of the Sciara del Fuoco (right side) contain blocks of lava and flows (basaltic), localized in a down-dropped portion of the volcano flank that may represent a sector-graben. Photo: F. Bullard, from *Volcanoes, in History, in Theory, in Eruption,* Univ. of Texas Press, 1962.

PLATE 35A

The stratovolcano Vulcano, Lipari Islands. The base of the complex of Vulcano is elliptical; the north-south axis is 15–20 kilometers long; the east-west one is 10–12 kilometers long. Off the northern coast is a small extension, Vulcanello, a 1,300-meter diameter, 123-meter high cone joined to the mainland (by an eruption in 1550) by volcanic ash. The cone of Vulcanello is 500 meters above sea level. Much of it is relatively flat; this portion of the peak is called the Piano (Plain). The Piano gradually declines north-ward and becomes bow-shaped. Its crescent-shaped rims enclose a large amphitheater of the Great Crater of Cono di Vulcano, which is 100 meters deep. Part of the crater rim is cultivated. At the north-east base of the cone some small lava projections (Pietre Nere) reach the sea. The cone slope angle is about 35 degrees and is made up of lavas only in the lower portions. Most of the volcano is made of tuffaceous material. The earliest effusive products were basaltic lavas, the later ones trachyandesitic. The activity of Vulcano is characterized by successions of short eruptive bursts separated by fumarolic and solfatraric activity. The last lava flow was in 1771, the last eruption in 1890. Photo: Italian Air Force.

PLATE 35B

View from the southern tip of Lipari south to the Island of Vulcano. In the background, the amphi-theater surrounding the Great Crater can be distinguished. The highest point (Mount Aria) on the southeast edge of the Island is 500 meters above sea level. The eastern edge of the Island is Punta S. Lucia; the western edge (on the right) is Mount Lentia. The top of Mount Saraceno is also visible. Vulcanello is in the left foreground. Center foreground shows an elliptical crater (Forgia Vecchia) on the north slope of Vulcano. The crater of Forgia Vecchia is 150 meters above sea level. A wall divides the crater into two parts, an upper and lower one. Lava flows are shown near the base of the volcanic complex on the right. Photo: Geophys. Lab., Carnegie Inst. Washington, D.C.

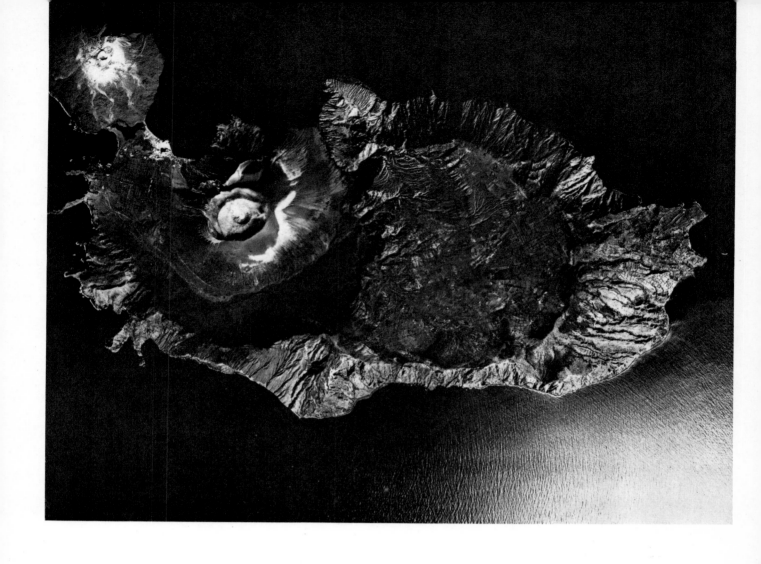

PLATE 36A

Aerial photo showing the snow-covered summit cone with several central craters, and numerous parasitic vents on the flanks of Etna, an active volcano near the eastern coast of Sicily. This volcano, almost 30 kilometers wide at sea level, rises to an altitude of 3,240 meters. It has been described as a basalt dome, surmounted by a smaller stratovolcano of trachyandesite composition. The smaller, partial crater lies along the northeast rim of the major vent crater which contains a still smaller nested crater. Several adventive cones can be seen on the flanks of Etna. These flanks are marked both by lava channels, from numerous flank eruptions, and by erosion furrows and ridges. A linear explosion pit zone appears on the eastern slope of Etna; a line of explosion craters, running north-northeast, is visible within the central crater. Photo: Italian Air Force.

PLATE 36B

View looking west at N.E. slopes of Mount Etna. Although the overall shape of the cone of Etna is simple, there are many details of structure due to a complicated history of eruptions. For example, the Valle del Bove, considered to a major collapse feature, extends along the entire ENE slope; the valley is about 5 kilometers wide with wall, reaching a maximum height (on the west) of 1,200 meters. Etna, like Vulcano, is characterized by a prevalence of lavas over pyroclastics at the base. The percentage of pyroclastics increases with altitude. However, the photo shows some recent lava flows along the upper slopes. The entire edifice of Etna has been built up by the activities of two eruptive centers, similar to Fujiyama. The older one, the volcano of Trifoglietto, collapsed to form the Valle del Bove at the end of its activity. The younger one, the presently active center, is the volcano of Gibello from which steam may be seen escaping. Photo: Geophys. Lab. Carnegie Inst. Washington, D.C.

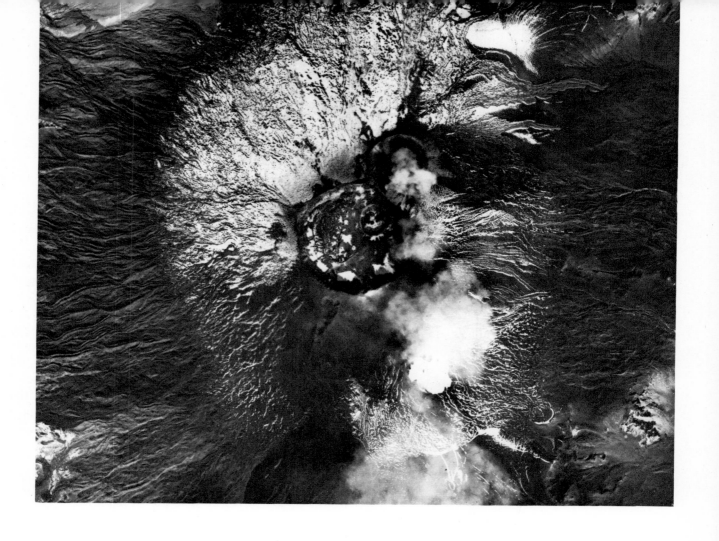

PLATE 37A

The Ring Complex of Ardnamurchan in northwest Scotland. On this peninsula, just north of the Island of Mull, the roots of a Tertiary caldera are exposed in the barren hills and shores. The pronounced circularity of the structural framework of the caldera base, defined by ring dikes, cone sheets, and intrusion centers, is evident in this aerial photo. South is at the bottom and west is on the left; the length of the photo is equivalent to approximately 11 kilometers. A generalized geologic map of the Ring Complex appears on page 14 of this atlas. With the aid of this map, it will be possible to pick out the three major eruptive centers, the steep, outward to inward dipping ring dikes, and the lower angle, inward dipping cone sheets. The rocks of the intrusive centers are mainly gabbroic, with some trachytic differentiates. The ring dikes vary in rock types from tholeiitic and alkali basalts to eucrites and quartz-dolerites. Most cone sheets are quartz-dolerites. Agglomerates, tuff beds, and pitchstone of trachytic, rhyolitic and dacitic composition are associated with some of the recognizable vents in the eruptive centers. The volcanic structure is surrounded by Moine schists, Triassic and Jurassic sedimentary rocks, and Tertiary flow basalts. Photo: Crown Copyright Geol. Surv. Photo. Reproduced by permission of the Comptroller, H.M. Stationery Office.

PLATE 37B

Mount Pawtuckaway, in Rockingham, New Hampshire, a nearly circular ring–dike seen in an aerial photo. The Pawtuckaway Mountains are a group of hills about 3 kilometers in diameter that rise 100–160 meters above the surrounding lowlands. The mountains are composed dominantly of a core of dioritic rocks that underlie valleys surrounded by a ring–dike of coarse-grained monzonite that forms the highest hills. The adjacent lowlands are underlain by quartz monzonite. The southwestern segment of the mountains is low because the ring–dike is cut off along a northwest-trending high-angle fault. Photo width covers 6.2 kilometers. Caption and Photo: U.S. Geol. Surv. Prof. Paper 590.

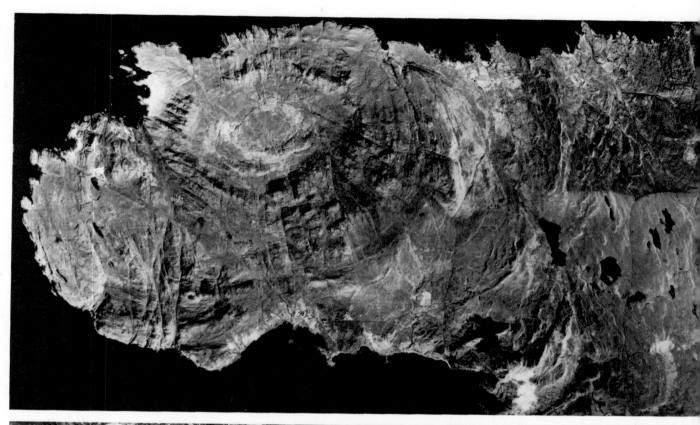

PLATE 38A

Broken Top, viewed from the southeast. Broken Top is a glacially eroded andesitic stratovolcano in Deschutes County, near Bend, Oregon. The interior cirque shows freshly exposed beds of pyroclastics and lava flows indicative of the many volcanic episodes responsible for the formation of the volcano. The Three Sisters, likewise andesitic stratovolcanoes, appear behind and to the left of Broken Top. Photo: Oregon Dept. Geol. Min. Ind.

PLATE 38B

Active volcanism in the Aso Caldera, Kyushu, Japan. The Aso Caldera is one of the largest Calderas in the world an with area of 3,600 square kilometers; the highest elevation is 1,592 meters. The active volcano in the east-west chain on the caldera floor is Mount Naka-Dake within which Aso-san is erupting. The caldera has had continuous activity on its floor since the 13th century. More than 10 volcanoes involving 50 craters constitute the interior cone complex. The caldera has a rather complicated outline showing several semicircular embayments. This feature is strongly indicative of the caldera having been formed by many circular collapse units. The basement of the Aso caldera is made up of Paleozoic, Mesozoic and Tertiary sediments together with granite and granite porphyry. From late Tertiary to early Pleistocene, eruptions of olivine-pyroxene andesite and pyroxene-hornblende andesite took place to form the volcanoes. In the latter part of the Pleistocene, about 175 cubic kilometers of pumice was erupted on the caldera flanks as an ash flow which filled valleys as far as the western and eastern coasts of Kyushu. Photo: Japan Air Lines.

PLATE 39A

Inside view of the central crater of the cone of Vesuvius (within Monte Somma) showing the structure of the walls three years after the 1906 eruption. The cone was build up from pyroclastic deposits of tephritic leucitite through a series of explosive eruptions. Dikes, sills, and lava flows are interspersed with the fragmentary deposits that make up the stratified layers of ash, lapilli and bombs which comprise the bulk of the materials within this stratocone. Some of the pyroclastic units laid down during single eruption periods are 10 or more meters in thickness. The rim of Monte Somma appears to the north. Photo: Geophys. Lab. Carnegie Inst., Washington, D.C.

PLATE 39B

Part of the inner wall of Monte Somma at Vesuvius, near Naples, Italy. This volcanic structure was built up from successive eruptions of explosively-ejected pyroclastic materials, largely phonolites and leucitites, interspersed with effusions of lavas. Layers of ash and lapilli, mixed with lava flows, are shown here dipping generally outward from the enlarged central cone. Dikes and sills were intruded across or between the layer units. In the foreground are fumarole mounds (whose shapes are somewhat modified by water erosion) produced during the volcanic activity of 1906. Photo: Geophys. Lab. Carnegie Inst., Washington, D.C.

PLATE 40A

Exposure by postglacial stream erosion of a lava cone in the Sveinar crater row, along the east wall of Jokulsa canyon, east of Lake Myvatn, in northern Iceland. The Sveinar crater row developed about 6,000 years ago during fissure eruptions. Here, the NNE fissure has intersected the older N-S Sveinar graben, a structure some 30 kilometers long, 0.5 kilometer wide, with vertical displacements less than 20 meters. The lava cone, with its prominent vertical feeder dike, is about 40 meters high at its vent. It has built up by lava outpourings of tholeiitic basalt that spread from the vent and pile up as spatter (near top of dissected structure). The cone rests on an erosion surface (prominent line near top of feeder) on an older lava flow with well-developed columnar jointing. This flow, about 25 meters thick, overlies another flow of similar thickness, at the base of which is a thin sedimentary layer of sand and gravel Photo: S. Thorarinsson.

PLATE 40B

Oblique cross section through part of a tuff cone (possibly a littoral cone) at Karl's Bay, northeast James Island, Galapagos, as exposed by wave erosion. Caption and Photo: T. Simkin.

PLATE 41A

Sheveluch, a compound volcano in central Kamchatka. The volcano is located at the intersection of the northeastern Kamchatka and the northwestern Aleutian tectonic zones. The northeast part of the structure is an ancient dome-shaped volcano built up of thick (> 100 meters) flows of hornblende andesite overlying pyroclastics of a more mafic composition. The southwestern portion is a relatively low pre-glacial stratovolcano made up of pyroxene andesites and cut by dikes of andesite basalts and basalts. An erosion caldera, 3 kilometers in diameter, formed at the summit of this structure, and in historic time eruptions occurred within it forming hornblende andesite domes. Directed volcanic blasts, involving nuées ardentes, occurred in 1944–1950. Adventive cones occur in the western and southern slopes of the southwestern portion of the complex. Photo: G. S. Gorshkov

PLATE 41B

Prevo Peak Crater, central Simushir Island, Kuriles, Soviet Union. The volcano, 1,360 meters above sea level, is a truncated cone with a base diameter of 6 kilometers. The upper part of the cone is relatively little dissected; only along the southwestern side does a large gulch occur. Two small adventive cones are at the foot of the western slope. The summit crater has a diameter of 600 meters. Within the crater is an internal cone with a deep funnel-shaped crater containing a vertical-walled pit crater. The diameter of the interior crater is 350 meters, that of the pit crater 200 meters. The last eruptions in 1914 were explosive (Vulcanian). The cone is made up of mafic andesites. Photo: G. S. Gorshkov.

PLATE 41C

Talbachik caldera complex, central Kamchatka, Siberia. Two calderas occur here; the older and extinct Ostry Tabalchik is shown on the left and the active caldera Plosky Tabalchik on the right. Plosky Tabalchik is 3,085 meters above sea level and is composed of andesite basalt. The active caldera is 4 kilometers in diameter on the floor of which (to the west) is a 2-kilometer active crater, the bottom being covered with ropy lava. Within the western portion of the 2-kilometer crater is a craterpit 300 meters in diameter and over 150 meters deep. Quantities of solfataric gases are emitted from this pit crater. A lateral eruption occurred in 1951 producing flows of olivine augite basalt. Large (3 cm) zoned basic plagioclase crystals occur in the lava. Lateral cone alignments occur on the southern and northeastern slopes of the volcanic complex. The type of eruption is in between that of Hawaii and Etna. The entire structure is at the intersection of an SSW fissure and regional NNE fracture zone. Photo: G. S. Gorshkov.

PLATE 41D

Tiatia stratovolcano, northeast Kunashir Island, Kurile Islands. The volcano has a well-developed somma within which a central cone projects 400 meters above the somma inner floor (atrio). The base diameter of the stratovolcano is 16–18 kilometers, that of the somma crater 2,550 meters, and that of the summit crater of the central cone 250 meters. The highest point is 1,822 meters above sea level. The southwestern slope of the central cone of Tiatia is built of andesite basalt; the other slopes are andesitic pyroclastics. The rim of the somma is well preserved and ruined only in the northwestern part. Note the lava streams in the southwestern part of the atrio. These lavas have descended from the summit crater of the central cone, filling the atrio up to the somma rim. Two small lava streams surmount the somma rim (at the lowest point on the northwest side) and descend along the outer slope of the somma. Photo: G. S. Gorshkov.

PLATE 42A

The crater lake at the summit of Ruapehu Volcano, North Island, New Zealand. This steaming lake has been the center of several spectacular eruptions since 1895; prior to this the volcano was thought to be extinct. In 1895 an eruption tossed lake water many hundreds of meters high, much like a gigantic geyser. In 1953, the upper wall enclosing the lake gave way, sending a torrent of water into a nearby valley; this weakened a railway bridge which collapsed shortly thereafter as a train crossed it, killing 153 persons. Photo: S. N. Beatus, New Zealand Geol. Surv.

PLATE 42B

View into the central crater of Anak Krakatau showing the hot crater lake, from which steam is produced and the layered ash units comprising the inner walls. The crater lake is 350 meters wide. Photo: Indonesian Geol. Surv., contributed by R. Decker.

PLATE 43A

View from the southwest looking into the active summit caldera of Mokuaweoweo on Mauna Loa volcanic complex, in Hawaii. This caldera is over 5 kilometers long and more than 2 kilometers wide. It joins with South Pit, the first of four smaller pit craters that extend towards the foreground of this aerial oblique photo. Many lava flows (most of the distinct ones were produced in the last 100 years) occur along the flanks of Mauna Loa. Those visible here are associated with fissures in the Southwest Rift Zone. Activity along this zone is evident in the photo, both in the main caldera and to the left of the pit craters. The lighter river-like, sinuous lava streams were still hot enough to flow at the time this 1940 photo was taken. Note that one stream is pouring as a "waterfall" into the pit crater in front of South Crater. On the horizon in the center is the volcanic complex of Mauna Kea. Photo: U.S. Air Force.

PLATE 43B

Vertical aerial photograph into Lua Hau, one of the pit craters located on the Southwest Rift Zone. Pit craters are characterized by almost vertical walls of basalt flows; no disruption of the flow margins at the pit edges occurs. Photo: U.S. Air Force.

PLATE 44A

Vertical aerial view of Cone Crater (left) and two small pit craters in the Kau Desert, Kilauea Volcano, Hawaii. These pits craters are about 40 meters wide and 80 meters deep. All three craters are possibly connected at depth by a large lava tube. Material erupted from Cone Crater built the small spatter cone and fed the pahoehoe flow that extends towards the right side of the photo. Caption and Photo: R.S. Fiske, U.S. Geol. Surv.

PLATE 44B

The Makaopuhi pit crater at the end of the road (in 1929) that follows the chain of craters extending southeast from Kilauea Island of Hawaii. This elongate crater, almost 1.5 kilometers in length, is actually a double pit crater. Successive collapses over a period of time have led to two general floor levels. The deeper pit on the right (west) was formed by further collapse of the pit on the left. In March, 1965, this pit crater was active again, with a lava lake filling most of the deeper pit (see Plate 104B). Photo: U.S. Air Force; Contributed by R. L. Smith.

PLATE 45A

Viti, a maar crater within the Askja Caldera, Iceland. Bedded pyroclastic deposits are present on the walls of this 130-meter diameter crater which was formed in 1724. The walls are extremely steep and the water filling the floor is warm. Viti plays an important role in the history of Askja caldera. In the late postglacial period, the Dyngjufjoll area was volcano-tectonically raised, probably due to the intrusion of a basaltic laccolith. Post-tectonic Askja lavas were then followed by subsidence of the top of the structure forming the Askja depression. Basaltic lava was produced from peripheral eruption centers at the northeastern and northern sides. In March 29, 1875, Viti was explosively formed producing 3–4 cubic kilometers of rhyolitic pumice; this was immediately followed by collapse of what is now the Knebel interior caldera. Basaltic eruptions have typified the caldera since 1921. Collapse of a caldera floor to form an eccentrically located caldera within it also occurred in Kilauea, but without pumice eruption. Photo: A. Patnesky, NASA.

PLATE 45B

Ljotipollur maar, southcentral Iceland. The maar is 1.5 × 0.75 kilometers, with long axis of crater in the northeast direction. Interior slope angles range between 25°–35°. Walls are of interbedded basalt flows and basaltic ash, some zones of which are highly ferruginous. Photo: J. Green.

PLATE 46A

The Mosenberg, one of a cluster of maars and tuff rings in the Eifel district, west of the Rhine graben, northwest of Bonn, Germany. The term "maar", meaning "crater-lake", originated from these explosively formed volcanic depressions in the Eifel Mountains. These Craters are the surface expressions of diatreme pipes (such as those of the Swabian pipe district east of the Black Forest in Germany.) Most of the Eifel maars now consist of low-rimmed (and forested) craters filled with water fed from groundwater. The pyroclastic deposits overlie nonvolcanic rocks, except in the feeder pipe zone. The structures seen here represent above-normal accumulations of expelled pyroclastics, so that they are almost cone-like in character. The volcanic activity in this region occurred about 12,000 years ago. Photo copyrighted by Hansa Luftbild BmgH, Münster Westf., West Germany.

PLATE 46B

Hole-in-the-Ground, Lake County, Oregon, a late Pleistocene maar, viewed from the south. This circular crater is about 1.6 kilometers wide and the highest point on the east rim is 153 meters above the crater floor. Basaltic flows, **not** associated with the formation of Hole-in-the-Ground, are exposed in the north wall and underlie explosion tuff breccias. A thin pumice layer mantles the surface. Alluvium caps a presumed thick ($>$ 500 meters) funnel-shaped layer of collapse debris and volcanic breccia on the crater floor. An older ash flow tuff is exposed on the crater wall just above the alluvium on the floor. White area on floor is a small playa. Flat Top, an eroded maar with a basalt cap, appears in the upper right. The structure is described by Peterson and Groh (1961). Photo: Oregon Dept. Geol. Min. Ind.

PLATE 47A

Flat Top, a rounded lava–capped mesa, Lake County, Oregon. As described by N. Peterson, this mesa is the remnant of a large tuff ring composed of inward-dipping thin layers of yellow-brown palagonite tuffs and breccias. After the explosive phases that formed the layered tuff ring, basaltic lava welled up inside to form a molten lake which overflowed the northwest rim. The cooling basalt provided a resistant cap to cause the landform to be inverted into its present position. Photo: Oregon Dept. Geol. Min. Ind.

PLATE 47B

Fort Rock, Lake County, Oregon. Remnants of the rims of this large basaltic tuff ring about 600 meters in diameter once projected as an island above ancient Fort Rock Lake. The present elevation of the rims is 100 meters. The steep wave-cut cliffs of Fort Rock display the initial dips of thin layers of palagonite tuff that accumulated when exploded debris settled around a broad crater. Wave action probably destroyed a possibly preweakened south wall of the ring. Near-by Hole-in-the-Ground was unaffected by wave action as it was above the level of the lake. Photo: Oregon Dept. Geol. Min. Ind.

PLATE 48A

Zuni Salt Lake, a maar crater 100 kilometers south of Gallup, New Mexico. The depression, about 1.5 kilometers wide is filled with saline water. In the low rim, dense thin basalt flows overlie Cretaceous sandstones. Veneering the basalt are deposits of coarse volcanic ash containing fragments of the underlying sediments. The two cinder cones on the floor of the crater are unusual occurrences for a maar crater which generally contains no internal volcanic features. The presence of the cones lends credence to the concept that the origin of maar craters may range from single explosions to multiple explosions and jetting processes. The hills in the background to the south are sedimentary. Photo: J. S. Shelton.

PLATE 48B

The Ubehebe Craters in northern Death Valley, California, examples of explosion craters or maars in which nonvolcanic bedrock is exposed in the walls. The larger crater on the left, about 730 meters across and 150 meters deep has sandstone and conglomerate exposed in the lower and middle parts of the steep crater walls; the upper part of the crater walls exhibits layered basaltic ash with sedimentary fragments. The larger craters are centered on a fault zone and were formed by explosive ejection of bedrock and fragmentary volcanic debris as gas-lava reached the vicinity of the ground surface, probably encountering water. The small crater to the right, which formed later, started as a spatter cone with a small basalt flow and finally ejected basaltic ash. Large volumes of dark basaltic ash and lapilli, mixed with fragments of sedimentary bedrock, were showered on the surrounding area from the main crater. This deposit is now undergoing "badlands" erosion, with steep-walled gullies and slopewash furrows cut into the soft pyroclastic materials. The intersecting, scalloped depressions around these maars are older explosion craters. Photo D. J. Roddy, U.S. Geol. Surv.

PLATE 49A

Crater Elegante, a nearly circular maar crater, Pinacate volcanic field, northern Mexico (Plate 53). The crater diameter is about 1,500 meters, its depth 243 meters. The lowest point on the floor is a little over 60 meters above sea level. Erosional remnants of three older cinder cones are surrounded by the rim ejecta of Elegante and a fourth is exposed in cross section on the southeastern part of its walls. A younger lava flow lies nearby on the west and southwest, and others are present at greater distances to the north. A composite dike of basalt porphyry about 70 centimeters in maximum thickness appears as a spine projecting from the basaltic cinders on the east wall of the maar. Gently outward–dipping basalt units, some vesicular, are exposed in the cliffs up to 8 meters thick on the crater walls. Post-dating the lava flows are scoria and basalt breccias. Over these are rim ejecta, about 50 meters thick, with a somewhat platy structure. The rim is rich in fragments of vesicular basalt. Photo: T. Nichols.

PLATE 49B

Cerro Colorado, a recent volcanic feature in the Pinacate Volcanic Field in northwestern Sonora, Mexico, a light reddish brown tuff ring that rises over a nearly flat surface some 24 kilometers northeast of Pinacate Peak. The tuff ring is about 1.6 kilometers in diameter and 110 meters high and surrounds an asymmetric basin about 850 meters by 1,050 meters in diameter. A young volcanic sequence consisting of tuff breccias forms the entire hill. The lowest unit is a pinkish, thinly-bedded ash with scattered lapilli and bombs. A gravelly tuff-breccia overlies this unit and contains granitic and metamorphic rocks. An angular unconformity exposed on the inner crater walls indicates repeated collapse of the inner depression. The hill is capped by a thick section of stratified tuff breccia similar to the lower unit, but more reddish brown. These soft tuff-breccias are subject to active erosion. One large mass of tuff-breccia, detached from the steep lower wall on the southeast side of the depression, has moved downward in very recent times. The center of the interior depression is a playa. Photo: D. Roddy, U.S. Geol. Surv.

PLATE 50A

Basaltic tuff rings, west tip of Ambrym Island, New Hebrides. (See Plate 17). The diameter of the largest ring shown is over one kilometer. Phreatic eruptions with or without sustained jetting of basaltic ash result when magma contacts the water table or near-surface water–saturated sediments. As observed at Auckland in New Zealand or the Kazinga Channel area in Uganda, a higher degree of explosivity producing maar-type craters is crudely associated with topography. Cinder cones predominate at higher elevations and maars at lower elevations. Overlap patterns are particularly well shown here. Photo: Hunting Surveys, Ltd., London; contributed by G.J.H. McCall and P.J. Stephenson.

PLATE 50B

Brown's Island, Waitemata Harbour, New Zealand, showing tuff ring (right), scoria mounds in crater (center), and lava flows (bottom left). Eastern slopes of Rangitoto Island (Holocene basalt) are visible in background. Caption and Photo: B. Thompson.

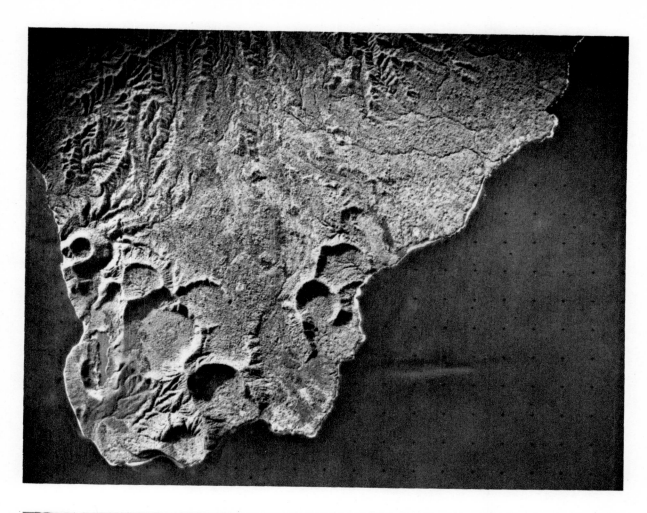

PLATE 51A

Diamond Head tuff cone, Oahu, Hawaii. This typical bowl-shaped crater is surrounded by bedded palagonite tuff with a maximum rim-to-rim diameter of one kilometer. Dips of the tuff (to 35 degrees) are anticlinal in form, coincident with the topographic rim except where it is modified by erosion. The combined topographic/structural asymmetry indicates that prevailing (toward southwest) trade winds were active during eruption. Such cones are common where ascending basaltic magma contacts water, resulting in steam-charged eruptions. Caption and Photo: R. V. Fisher, Univ. Calif., Santa Barbara.

PLATE 51B

Littoral cone complex, south shore, Island of Hawaii. Highest part of complex (80 meters) is named Puu Hou. It was built in only five days from hyaloclastic debris dispersed by steam explosions caused by lava entering the sea. Flows were from flank eruption of Mauna Loa in 1886. Debris, radially ejected from explosion foci, fell partly on land and partly at sea, resulting in crescent-shaped rims (half-cones) breached by the flows that built them. Layered basalt in the foreground is a prehistoric flow from Mauna Loa. Caption and photo: R. V. Fisher, Univ. Calif., Santa Barbara.

PLATE 52A

Some of the late stage adventive cones, accentuated by snow, along the upper flanks of the composite volcano, Mauna Kea, Hawaii. These cones are built mostly of spatter with relatively little cindery material as a result of local explosive fire-fountaining and some ash-producing eruptions. Photo: U.S. Air Force.

PLATE 52B

Tagus (top) and Beagle (bottom center) Craters, Isabela Island, Galapagos Islands. The craters, both about 2 kilometers in diameter, are composed of basaltic tuff and pyroclastics. Both craters have been eroded by wave action and Beagle Crater was later invaded by a recent basaltic lava flow. Note the rim furrows on Beagle Crater; these are caused by water erosion but were possibly initiated by fragment scouring during explosive eruptions. Photo: taken from altitude of 6,400 meters by the U.S. Air Force, released courtesy of the Government of Ecuador, and published with permission from Banfield, Behre, and St. Clair (1956).

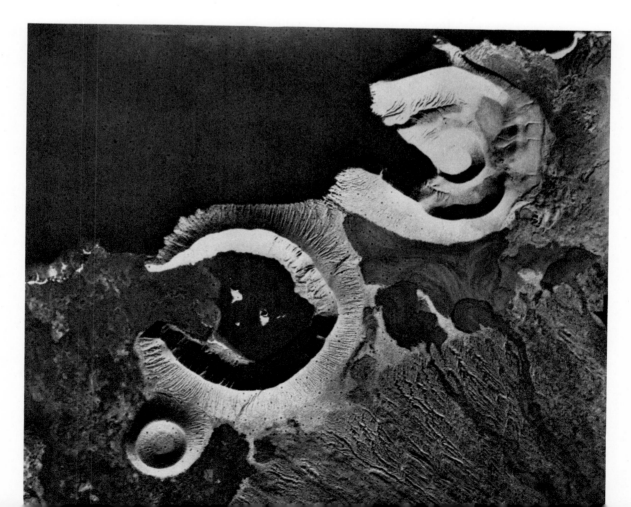

PLATE 53

A near vertical view of the Pinacate Volcanic Field, taken from the Gemini IV spacecraft in 1965 by E. H. White, II, over northwest Mexico. The Gulf of California is at the lower left; the area shown is in northwest Sonora, with part of southern Arizona at upper right. At the center of the picture is the Sierra del Pinacate, a recent volcanic field (the most recent eruptions having occurred only a few hundred years ago), measuring roughly 55 by 80 kilometers. The volcanoes are maars and cinder cones that have erupted flows and pyroclastics of basaltic composition. The photograph shows the overall structure of the Pinacate volcanic field and its relation to the regional structure. The linear ranges at the right are composed of Mesozoic igneous and metamorphic rocks block-faulted in a typical basin and range pattern. Although these northwest trending faults dominate the regional structure, they appear to have had little influence on the location of vents in the volcanic field, suggesting a deep subcrustal origin for the magmas. Orbital photography of other Quaternary basalt fields in North America and North Africa also do not show much correlation of the location of vents in the volcanic field with regional structure. Caption: P. Lowman. Photo: NASA.

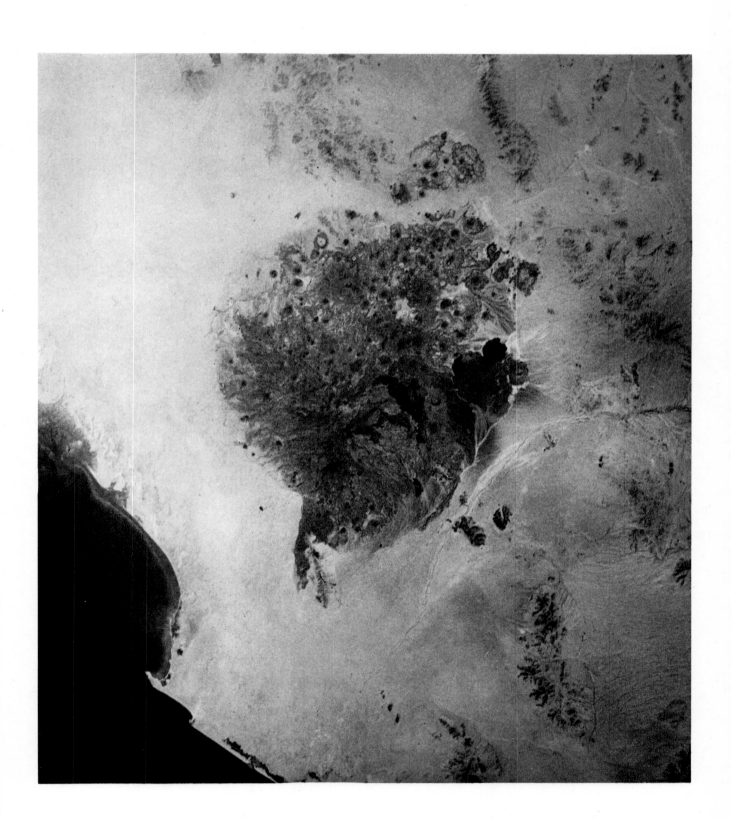

PLATE 54A

A cluster of cinder cones in the Auvergne district of France. The crater of the Puy de Pariou and its somma appear in the foreground. In the background (looking north) is a complex of cones and craters including the arcuate Puy de Gouttes and the Puy Chopine (center back). The Puy Chopine is a peléan spine, having uplifted a part of the granitic basement. The Auvergne region is famous as the Graveyard of the Neptunists. Studies in the 18th century by Guettard and Desmarest, and later by von Buch and others, demonstrated the true volcanic nature of the Puys and destroyed the concept of the Wernerian school that held igneous rocks, including volcanic flows, to be deposited as chemical precipitates. This way one of the key discoveries by which geology became a science. Most of the Puys are cinder cones associated with coulees or flows of basalt, but there are also extruisve trachytic domes built up by effusion of viscous, silicic lavas. Photo: published by permission of the Parc des Volcans, Loic Jahan, photographer; submitted by M. Derruau.

PLATE 54B

Lava Butte, 16 kilometers south of Bend, Oregon on Route 97. The cone is on the site of a northwest series of fractures. A vent on the southern side of the cone produced a large 26-square kilometer aa flow which streamed south and west and north. Much of the 153-meter high, 610-meter wide cone is made up of basaltic clinkers and cinders many of which are iridescent due to a thin filming of iron oxide. An observatory and historical museum are located at the summit which is accessible by automobile. Photo: H. Coombs, Oregon Dept. Geol. Min. Ind.

PLATE 55A

Cinder cone and spatter cones along the Laki fissure (gja) in Iceland. The foreground and background show a small portion of the 12 cubic kilometers of basaltic lava produced from the 25-kilometer long fissure in 1783. The ash and fumes from this eruption caused crop failures which resulted in a famine taking over 10,000 lives. Caption and photo: R. Decker.

PLATE 55B

Oblique aerial view of 1940 spatter and cinder cone on the floor of Mokuaweoweo Crater, Mauna Loa Volcano, Hawaii. A fissure that formed during the eruption of 1949 cut the cone and showered spatter on it. Caption and Photo: R. S. Fiske, U.S. Geol. Surv.

PLATE 56A

General view to the northwest of the San Francisco Mountain Volcanic Field northeast of Flagstaff, Arizona. O'Leary Peak, a large dacite dome, appears on the far left horizon; beyond it and to the right is the north-northeast trending group of cinder cones and maars. The second group, which trends generally northwest, occupies most of the photo. In the foreground is an older maar, Moon Crater, about 1.8 kilometers in diameter. Its broad, low rim and large central depression distinguish it from typical cinder cones which have steeper sides and a narrow summit crater. This maar contains a well-developed central peak which is a pitless cone. Caption and Photo: J. McCauley. U.S. Geol. Surv. Astrogeol. Branch.

PLATE 56B

Vertical aerial photo of the S.P. Crater and its associated lava flow near the north end of the northern group of cinder cones of the San Francisco Mountain Volcanic Field, close to Flagstaff, Arizona. The basalt flow issuing from the northwest base of the cinder cone (itself about 0.8 kilometer wide at the base) extends northward for more than 5 kilometers. The light-colored mottled areas and patches, mainly around the edges of the flow, are deposits of wind-blown materials. Photo: T. Nichols.

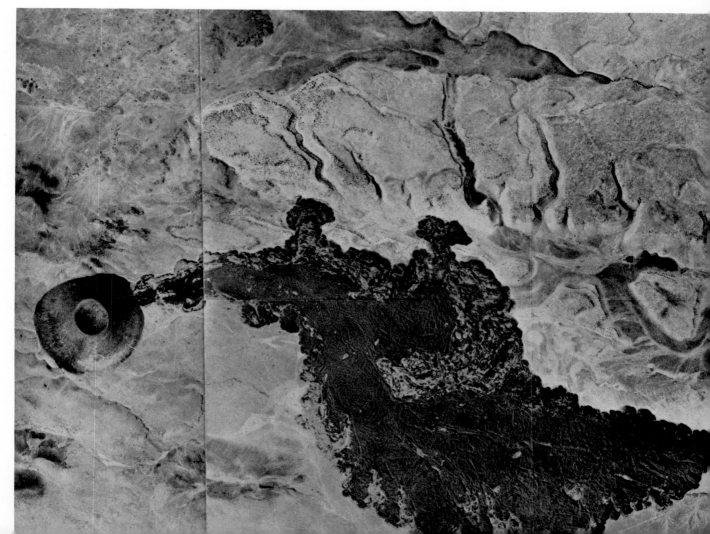

PLATE 57A

Sunset Crater, a typical basaltic cinder cone nearly 330 meters high and 1,600 meters wide at the base. View from the west. A considerable amount of hydrothermal alteration is present on the interior of the rim; both sulfur and gypsum are present. The exterior slope is 32 degrees. Slabby to platey aa lava covers much of the terrain to the west (lower). A small hill, Cinder Hill, lies to the north. The cone was formed in 1065 A.D., mantling the area with volcanic ash and cinders. The meadow in the foreground was formed in this way. Photo: T. Nichols. Inset shows an aerial photograph of this crater oriented so that the slopes in the lower part of the picture correspond with those shown in the foreground of the ground photo. Photo: U.S. Geol. Surv., Topographic Branch.

PLATE 57B

Radar image of a portion of the eastern part of the San Francisco Mountain Volcanic Field in northern Arizona. Note the rectilinear pattern of faults. Among the named cinder cones are (A) Sunset Crater, (B) Lennox Crater, and (C) Robinson Crater. Crater Number 81 (D) is one of the largest cones in the field. O'Leary Peak (E) is a silicic dome (see Plate 63A). The eastern end of the San Francisco Peaks acidic rocks is visible at (F). The Bonita Flow (G), coming from Sunset Crater, is barely visible. The front of a major flow unit (H), comprised of coalescing individual flows closely spaced in time of extrusion, is evident in the northern end of this part of the field. H. S. Colton has subdivided the history of the field into five stages of which the formation of Sunset Crater and its lava flows are among the youngest events recognized. Photo: J. McCauley, U.S. Geol. Surv. Astrogeol. Branch.

PLATE 58A

The Janus craters, San Francisco Volcanic Field in northern Arizona. A symmetrical double cinder cone has been produced by simultaneous or successive eruptions from closely spaced single vents. Structure is approximately 1 × 2 kilometers and about 160 meters high. Caption and Photo: J. McCauley, U.S. Geol. Surv. Astrogeol. Branch.

PLATE 58B

A linear pyroclastic cone 1.0 by 1.6 kilometers by 165 meters high, typical of many of the volcanic vents in the northwest part of the San Francisco field near Flagstaff, Ariz. The elongation results from simultaneous eruption along a fissure. Caption and Photo: J. McCauley. U.S. Geol. Surv. Astrogeol. Branch.

PLATE 59A

Herdubreid, a flat-topped volcanic mountain in northern Iceland. Viewed from the west, Herdubreid is one of the most prominent of all table mountains; its summit at 1,500 meters stands 1,000 meters above the surrounding terrain. This landform is a relatively narrow pillar of loosely cemented palagonitic rocks armored in part by columnar basalts exposed in the lower slopes. The basalt is petrographically similar to the palagonite breccias, both being porphyritic basalts. The table top is crested by a postglacial volcanic cone with a summit crater that rises 150 meters above the plateau. The plateau of Herdubreid is covered by a series of subaerially extruded lavas several tens of meters thick. A table mountain, and its companion form, the serrate ridge (Plate 59B), are generally considered to result from extrusion (or "intrusion") of magma within glacial ice. During this subglacial invasion, the material is molded against the walls of large voids melted into the ice. The piled-up basalt forms either elongate to equant mountains or ridges depending on the shape of the vent(s) supplying the lavas. The capping of a table mountain usually represents lava flows extruded subaerially after the ice sheet is completely penetrated. Photo: J. Green.

PLATE 59B

Kálfstindar, a linear volcanic ridge formed of hyaloclastite breccia (palagonite) of the Moberg formation, Sudvesturland, 10 kilometers east of Thingvellir on the southern margin of Iceland's central upland. This serrate ridge—the product of subglacial fissure or linear eruptions—is a characteristic and easily identifiable feature of the skyline of the Reykjanes-Langjökull zone. Kálfstindar trends northeast, parallel to the general trend of other eruptive and dilation fissures in southern Iceland. The highly brecciated, rough texture of the hyaloclastite, which, as an eruption product, never broke through the top surface of the ice sheet, is the result of rapid cooling and the pressure of overlying ice during the eruptive stage. Caption and Photo: J. D. Friedman, U.S. Geol. Surv.

PLATE 60A

Panoramic view of Usu composite volcano, Hokkaido, Japan. This stratovolcano stands on the southern rim of the caldera now occupied by Toya Lake and is composed of basalt and basaltic andesite. There are two domes on its summit, O-Usu and Ko-Usu. In 1910 a cryptodome formed on the northern foot of the volcano and from 1943 to 1945, a new dome, "Showa-Shinzan", composed of hypersthene dacite erupted on the eastern foot of the volcano as a parasitic cone. O-Usu (center) and Showa-Shinzan (left center) are shown in relation to the entire structure. Inset shows the lava dome of O-Usu composed of hypersthene dacite formed at the top of Usu volcano in Hokkaido, Japan. The dome reaches an altitude of 725 meters above sea level. Captions: K. Yagi, Photos: Y. Oba.

PLATE 60B

The lava dome of Showa-Shinzan formed by the extrusion of almost solidified dacite lava during the activity of 1944–45 at the eastern foot of Usu Volcano, Hokkaido, Japan. Caption: K. Yagi. Photo: Y. Oba.

PLATE 61A

The plug dome of Lassen Peak, seen from the southwestern slope of Lassen Peak, California. This dacite dome may be a somewhat cylindrical filling of an older vent or crater. The dome possibly has been lifted as much as 800 meters. The cliffs of the south face are striated and polished in a manner similar to the face of the Pelée spine (Plate 69A). Photo: J. S. Diller, U.S. Geol. Surv.

PLATE 61B

A plug of andesitic lava thrust up around May 19, 1915 during the major eruption period of Lassen Peak, in the California Cascades. This dome-like body resulted from upheaval of the floor of the explosion crater that had expanded during the previous year. The exposed plug was solid though hot at the time of extrusion, but viscous lava beneath it may have aided in pushing the blackish, dense glassy mass upward. The craggy, chaotic surface results from blocks breaking loose along cooling fractures. The white areas are pockets of snow. Photo: Geophys. Lab. Carnegie Inst. Washington, D.C.

PLATE 62A

Tholoid (dome) of the Bezymianny crater following the violent eruption of March 30, 1956. This explosion also tore away the southeastern slope of the volcano reducing its height by 200 meters, i.e. to about 2,800 meters above sea level. The energy of the explosion was about 10^{23} ergs. The composition of the plug lavas are hornblende and pyroxene andesites. Photo: G. S. Gorshkov.

PLATE 62B

Novarupta, a tholoid within a tephra ring, Valley of Ten Thousand Smokes, Alaska. The 65-meter high, 400-meter wide tholoid consists of large angular blocks of dacite resulting from the slow upward movement of viscous magma following the evolution of enormous quantities of a gas-glass-dust emulsion filling the Valley of Ten Thousand Smokes in 1912 (See Plate 101). The crater rim surrounding the tholoid is composed of dacite tuff and pumice fragments. A number of hydrothermal vents occur on the west side of the crater on the rim. Most of the alteration is of the sulfur-acid type as distinct from the halogen-acid alteration which is more typical of the fumarolic activity at the distal end of the valley flows. See Lovering (1956). Photo: A. H. Chidester, U.S. Geol. Surv.

PLATE 63A

O'Leary Peak, a complex extrusive dome of dacitic to rhyolitic composition, approximately 5 by 5 kilometers and 650 meters high, within the San Francisco volcanic field near Flagstaff, Arizona. Note the sill on the right. Several small cinder cones appear in the foreground. Caption and Photo: J. McCauley, U.S. Geol. Surv. Astrogeol. Branch.

PLATE 63B

Rhyolite lava dome and flows in Valles Caldera, New Mexico, shown in an aerial vertical photograph. The broad elongate mountain near the center is one of 10 similar rhyolite dome-flows that form a discontinuous ring within the Valles Caldera in the Jemez Mountains (see Plate 16). The conspicuous arcuate ridges on the south side of the mountain are ribs of vertically flow-banded rhyolite which have been accentuated by erosion of intervening zones of less resistant flow breccia. These semi-concentric ridges are part of a thick rhyolite flow that forms the base of the dome-flow complex and extends 12 kilometers to the west. Capping this basal flow is another dome-flow which originates at the grass-covered crest and extends 3 kilometers to the northeast; this uppermost flow also displays conspicuous concentric flow ridges. The small tree-covered knoll immediately northeast (upper right) of the dome-flow complex is a small isolated rhyolite dome. The Valles Caldera measures 19.2 by 22.4 kilometers. Caption and Photo: U.S. Geol. Surv. Prof. Paper 590.

PLATE 64A

Glass Mountain obsidian flows and domes in Siskiyou County, California. This aerial photograph shows that Glass Mountain consists of two obsidian flows, the younger of which runs northeast (towards upper right) from a summit dome. Flow structure and steep margins of flows stand out. The older Hoffman flow lies to the west and supports a moderate growth of pines. Its vent, Mount Hoffman, is outside the area of the photograph. The white patches on the Hoffman flow are pumice that has filled depressions on the flow surface. A line of small domes trends N 30° W from Glass Mountain. The flows are probably less than 1,000 years old. Caption and Photo: U.S. Geol. Surv. Prof. Paper 590.

PLATE 64B

Glass Mountain, Siskiyou County, California. This recent obsidian flow culminated the volcanic activity of the Medicine Lake Highland and is on the east rim of the Medicine Lake Caldera. The craters, some as much as 40 meters in diameter, may represent explosion features, perhaps formed as the viscous lava moved over wet ground. Caption and Photo: R. Greeley.

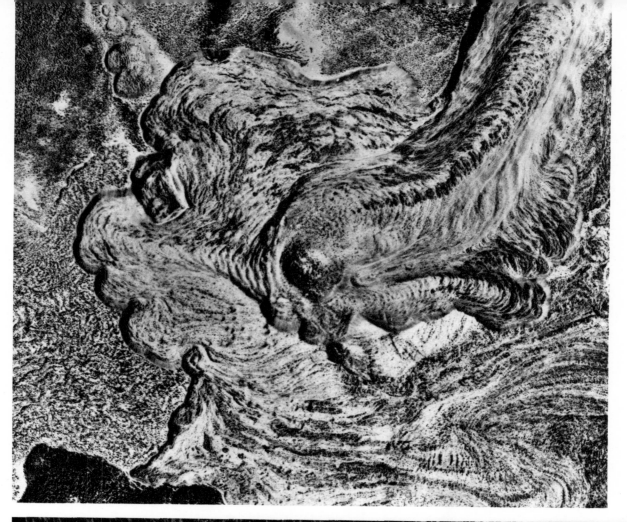

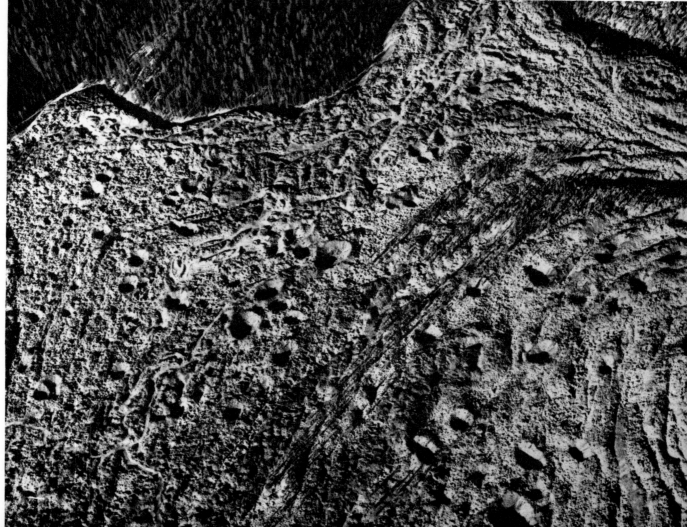

PLATE 65A

Aerial (vertical) photograph of Mono Craters in eastern California. This chain of rhyolite domes and flows has been described as citadel-like structural features whose nearly vertical walls are girdled with a continuous mantle of blocky talus and surmounted by a rampart of obsidian pinnacles. Glacial moraines are conspicuous in the southwest (lower left) quadrant of the photo. Photo width covers 7.5 kilometers. Caption and Photo: U.S. Geol. Surv. Prof. Paper 590.

PLATE 65B

Aerial oblique view towards the south (eastern front of Sierra Nevada Mountains in background) of the domes, coulees and tephra rings of Mono Craters, near Bishop, California. The Panum Crater (see Plate 66) appears in the center foreground behind Mono Lake. The Northern Coulee, a viscous rhyolite flow extends to the left of the line of domes; the frontal lobe of the Southern Coulee is seen in the background to the right. Between these coulees is the line of pumiceous ash-covered domes that trend generally southward. The Mono Craters lie along the same lineament as the Inyo Craters to the south. More than 20 domes are found in this system, which formed from spasmodic but relatively short-lived volcanic events about 5,000 years ago. Photo: R. von Huene, U.S. Geol. Surv.

PLATE 66A

The Mono Craters area as seen by high resolution side-looking radar (viewed to the north at top, with east to the right; the hills at the left [west] have experienced slant range distortion) along track (horizontal). The radar image clearly defines a series of cumulo-domes (including Panum at the top right near Mono Lake) some with central depressions and the steep-fronted Northern and Southern Coulees or rhyolite flows (compare with Plate 65). Photo length (top to bottom) covers 24.7 kilometers. Photo: NASA Earth Resources Survey Program and Allen Kover.

PLATE 66B

Thermographic infrared imagery of the main rhyolite-obsidian complex at Mono Craters, California. Thermal emission recorded over 8–14 μ wavelength region, at 8:00 a.m., June 4, 1965. Major effect in early daytime imagery is the enhancement of topographic and microrelief features subject to the most direct solar heating. Surface emission in the 8–14 μ interval depends on intensity and inclination of the incident radiant flux, and total absorptivity of the surface as well as thermal inertia of surface materials. Thermal shadows result from relatively low emission from cooler shaded areas. Note (a) outlines and texture of slopewash and fan deposits west of crater complex; (b) texture and lineaments in pumiceous tephra east of crater complex; (c) clastic cones and explosion pits of main range; (c_1) late-stage explosion pit in bedrock of Punchbowl; (c_2) tholoid completely enclosed by older tephra ring; (d) flow ridges and outflow centers in rhyolite-obsidian tholoids and coulees; (d_1) vertical turreted tholoid of Wilson Butte; (d_2) biscuit-shaped obsidian-rich Glass Mountain tholoid with prominent flow ridges and outflow centers; (e) outcrop pattern of June Lake basalt sheet west of crater complex; (f) configuration of Tahoe-age morainic ridges near Aeolian Buttes and South Coulee; (g) Aeolian Buttes inliers of quartz monzonite ringed by upper units of Bishop tuff; (h) wet spot (dark) in center of pumiceous sand flat east of Punchbowl, suggestive of playa conditions. See Friedman (1968). Caption and Image: J. D. Friedman, U.S. Geol. Surv.

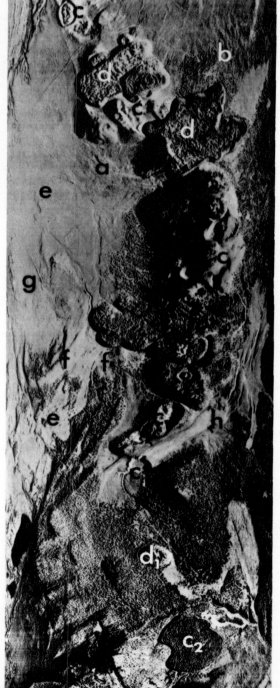

PLATE 67A

Howard Mesa in the San Francisco Volcanic Field northeast of Flagstaff, Ariz. This structure forms a domical mesa some 5 by 5 kilometers by 100 meters in dimensions, and is of intermediate composition. Lavas erupted from a central vent, but there are no pyroclastics associated with this volcano. This type of structure is believed similar to low domes observed in the Marius Hills region of Oceanus Procellarum of the Moon. Caption and Photo: J. McCauley, U.S. Geol. Surv. Astrogeol. Branch.

PLATE 67B

View to the southwest showing the region of the Air Mountains (also known as the Air ou Azbine) in Nigeria, taken from the Gemini VI spacecraft in 1965. The Air Mountains are part of a large Precambrian massif similar to the Ahaggar, 300 kilometers to the north, and other massifs of North Africa. Although relatively unknown because of their remote location, the Air Mountains include one of the world's most extensive ring-dike assemblages. The series of ring-dikes, stocks, and batholiths shown here is over 550 kilometers long in the north-south direction. Some of the individual intrusions, such as Monts Tamgak (the largest, at lower right) are over 30 kilometers wide. These intrusions are late-Precambrian and post-tectonic, and provide remarkably good examples of discordant batholiths. The major rock types in this area are granites, syenites, and rhyolites (of the alkaline clan), some containing tin, tungsten, and niobium minerals. Caption: P. Lowman. Photo: NASA.

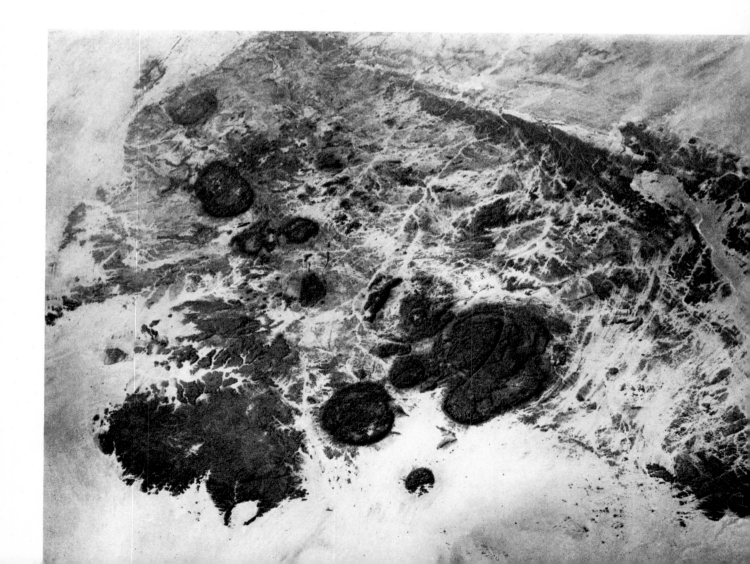

PLATE 68A

Bear Butte, a denuded laccolith near Sturgis, South Dakota. This butte, about 382 meters high, is adjoined to the east by a basin about 3 kilometers in diameter rimmed by Lakota-Falls River sandstone. The butte is composed of trachyte of which the talus slopes are most spectacular. Conspicuous silicification of upturned Paleozoic formations occurs on the flanks of the butte, together with much peripheral brecciation. The feature was formed in Tertiary time, conceivably contemporaneous with the rhyolites of the Lead district. Photo: South Dakota Dept. Highways.

PLATE 68B

Phonolite laccolith exposed at Traprain Law, East Lothian, Scotland. The laccolith is intruded into Calciferous Sandstone of Carboniferous age, outcropping at the base of the intrusive. Photo: B483-5, Crown copyright Geol. Surv. Reproduced by permission of the Comptroller, H. M. Stationery Office.

PLATE 69A

Two views of the great spine or "aiguille" of Mount Pelée, on the Island of Martinique in the West Indies. This spine, of silicic rock, presumably was extruded in a semi–rigid state from cracks in the cumulo dome of 1902–1903 in the crater from which earlier nuées ardentes had devastated parts of the island. (See Plate 3A). The spine began to form in November 1902, reaching a maximum height near 350 meters, but by mid-1903 had collapsed to its base. The left photo, looking from the east, shows the curved "back" of the spine on which vertical grooves were impressed during extrusion. The right photo, from the south, reveals the asymmetric shape of the spine. The rough, flat side may represent a fracture surface, suggesting that part of the spine remained in the dome. Photo: A. Lacroix.

PLATE 69B

Vent filling or "spine" of Mount Washington, Cascade Range, Oregon. The rock type in the vent filling, andesite, is the same as the enclosing stratovolcano rocks. The landmark owes its prominence to glaciation and erosion and not to a great upward thrust during the eruption in the sense of a Peléan spine. However, that some vertical movements were involved is recorded in the vertical striations still preserved in some areas on the base of the spine. Photo: Oregon Dept. Geol. Mineral Ind.

PLATE 70A

Eroded volcanoes of the Puy de Dome district, Auvergne, France. View to north–northwest from the 1,360-meter high Col de Guery. The Rochefort Valley is in the center. Two eroded necks of trachyandesitic volcanoes are shown, Roche Tuilière on the left and Sanadoire on the right. Sanadoire is 1,288 meters high; the distance between Sanadoire and Roche Tuilière is 950 meters. The chains of extinct volcanoes and domes of the Puys of Auvergne differ from the chains of cones at Threngslaborgir in Iceland in that the volcanoes and domes in France evolved during Tertiary and Quaternary time as the result of independent eruptions. In Iceland, the recent cones owe their origin to a single linear fissure eruption. Photo: J. Green.

PLATE 70B

A neck of volcanic breccia, the Rock of Saint Michel (left) and Le Puy en Veley, suburb of Aiguilhe, Auvergne District, France. Both of the hills shown are composed of indurated volcanic breccia which remained after the surrounding cones of softer pyroclastic material were eroded away. The building on the summit of St. Michel, built in the 10th century, is a chapel. Photo: French Cultural Services, New York.

PLATE 71A

Devil's Tower, banks of the Belle Fourche River, 40 kilometers northeast of Moorcraft, Wyoming. The top of the Tower is at an elevation of 1,560 meters, some 183 meters above its base, which is about the same distance the base is above the Belle Fourche River. The sediments forming the base include the Sundance and Gypsum Springs formations of Jurassic age overlying the Spearfish shales of Jurassic age; the sediments are only slightly deformed by faulting. The most impressive feature of the structure is the huge polygonal columns about 2 to 2.5 meters in diameter at the base and some 1.3 meters in diameter near the top. Note that the columns, mostly five-sided (but also four or six), flare outward at the base. The east-west diameter of the top is 55 meters, the north-south about 91. The rocks of the Tower are phonolite porphyry containing soda-rich orthoclase and augite with an outer zone of aegirite. Some apatite and magnetite are also present. Sedimentary strata, some 900 meters distant from the base of Devil's Tower, dip away about 2 to 3 degrees. On the other hand, the dip of the sediments near the Tower is inward from 3 to 5 degrees. This suggests that Devil's Tower is near the middle of a collapsed dome. Robinson (1956) suggests that the structure is an intrusive igneous mass which was never larger than its base and which is connected at depth ($<$300 meters) to a sill or laccolith. The orientation of the columns according to this hypothesis would relate to the cooling history of the parent mass. Photo: G. A. Grant, Nat. Park Serv. U.S. Dept. Interior.

PLATE 71B

The Ehi Goudroussou, one of the several volcanic necks in the northern and central Tibesti mountains of Chad in Africa. Trachyphonolitic magma consolidated in the vents of eroded volcanoes produces these isolated pinnacles of columnar lava. Note that the vertical jointing, which flairs out at the base, is similar in structural pattern to that exposed in the Devil's Tower, Wyoming. Caption and Photo: A. Pesce.

PLATE 72A

Ship Rock, northwest New Mexico. Ship Rock is a volcanic neck about 430 meters high and about as wide at the base. The mass is made up of a fine minette tuff-breccia admixed with fragments of sediment and plutonic rocks. Toward the top of the structure, there is inward dipping stratification due to the fallback of pyroclastics in the volcanic vent. Irregularly trending dikes traverse the breccia but are only a small part of the mass of the peak. Inclusions of granite and diorite are abundant in these pyroclastics. Radial and tangential dikes of monchiquite have been well exposed by erosion of hundreds of meters of sediments. Photo: J. S. Shelton.

PLATE 72B

Close-up of Ship Rock, New Mexico. The tuff-breccia composing the vent filling has been brought up from great depths as discussed by Williams (1936). Note the well-developed vertical jointing. Mancos shales (undisturbed) occur at the base of the monolith. Photo: J. S. Shelton.

PLATE 73A

The Hopi Buttes volcanic field, Navajo County, in northeast Arizona, looking west–northwest. Chezin Chota Butte in the foreground consists of volcanic ash beds capped by lava flows. In the middle background are the volcanic pipes or necks of Castle Butte and Red Check Butte. Near the horizon, left to right, are the Elephant, Pyramid, and Montezuma's Chair buttes which are also diatremes or volcanic necks in various degrees of erosional exposure. Photo: R. Sutton, U.S. Geol. Surv.

PLATE 73B

Agathla Peak, an eroded breccia pipe or diatreme west of Kayenta, northeast Arizona. This large tapering monolith rises over 300 meters above the conical pedestal of flat lying Chinle shales; its base is about 900 meters wide. Fully nine-tenths of the diatreme is made up of buff and gray tuff breccia, the matrix of which is pulverized minette and sediment breccias. Some breccia blocks measure over 7 meters in diameter although most are on the order a centimeter. Many of the plutonic inclusions are well rounded due to hydraulic churning associated with repeated explosions in the vent. A roughly synclinal structure may be observed on the summit of Agathla which may be attributed to fallback of tephra after volcanic explosions. Most of the dikes of minette cutting the pyroclastics of Agathla follow zig-zag courses, branching and swelling at random. Most dikes are on the order of 3 meters wide. A description of available stereo vertical angle photography of this peak is given by Denny et al. (1968, p. 14). Photo: J. S. Shelton.

PLATE 74A

Exposure along a hillside of the East Sturgeon Island diatreme or breccia pipe (lighter colored area) in Montana. Diatremes are elongate (generally with roughly circular cross section) pipes or conduits filled with fragmental material largely derived from the rock units through which the pipes pass; some volcanic rock types (commonly with unusual mineralogy and chemistry) are normally present as the enclosing matrix. The pipes are produced by explosive emplacement as fractures are widened and country rock broken off by gases formed by interaction between hot invading magma and groundwater. This phreatic process leads to a fluidization of the gas-magma-rock fragment mixture that circulates through the expanding vent. Some maars are presumed to be the surface expression of the escaping fluidized mixture. In desert country, the diatreme fill may be weathered out as necks or buttes (see Plate 73A) However, in more humid climates and/or where the diatreme material is not more resistant than its surrounding country rock, a diatreme may be more easily eroded than its surroundings and will have little or no surface expression. Thus, the diatreme shown here (and many of those in Missouri, Illinois, and Kansas) is recognized mainly by its lithologic contrast to the host rocks. Photo: C. Hearn, U.S. Geol. Surv.

PLATE 74B

Summit of Alhambra Rock, in San Juan County, Utah, part of a diatreme west of Mexican Hat, Utah. This breccia pipe is composed of minette (sanadine-biotite–bearing alkali basalt) together with fragments of country rock. Caption and Photo: T. McGetchin.

PLATE 75A

Differentially eroded dike of syenodiorite porphyry, Spanish Peaks, near Trinidad, Colorado. The vertical dike is about 10–20 meters above the surrounding terrain of sedimentary rocks. The well-developed, Eocene dike pattern of Spanish Peaks is not perfectly radial but has been influenced by a pre-existing system of joints developed during the early stages of folding of the Laramide orogeny (late Cretaceous). An analysis of the dike pattern in this area has been given by Badgley (1965, p. 325–327). Photo: J. S. Shelton.

PLATE 75B

Intersecting dikes northeast of Spanish Peaks, Huerfano County, Colorado. A large east-trending dike (upper half of photo) cuts across older radial dikes that trend in a southwesterly direction. The bedrock is mostly competent sedimentary rocks of late Paleozoic and Mesozoic ages. Dikes form prominent ridges that control the drainage pattern. Photo width covers 4 kilometers. Caption and Photo: U.S. Geol. Surv. Prof. Paper 590.

PLATE 76A

The inner wall of Monte Somma, surrounding the active Vesuvius volcano in southern Italy, viewed from the Atrio, showing both cross-cutting dikes and flows interlayered with bedded ash and breccia. In the foreground, the effects of over eight years of rain-induced erosion of ash from the 1906 eruption are evident. Photo: Geophys. Lab., Carnegie Inst., Washington, D.C.

PLATE 76B

Tuffaceous pyroclastics and basaltic flows exposed on the interior wall of Monte Somma, Vesuvius, Italy. The attitude of the bedding is gently outward dipping. Photo: Geophys. Lab., Carnegie Inst., Washington, D.C.

PLATE 77A

Dike complex at Puy de Sansy dome, Auvergne, France. The elevation of the Puy de Sansy is 1,886 meters; the original elevation of the Quaternary volcano at this site is estimated to have been 2,500 meters. Glaciation has eroded the structure exposing a classic array of dikes radiating from the main volcanic vent. The dikes are sansyites or leucocratic trachyandesites and trend N 55 W in this photograph; some are over 100 meters high. The Monts-Dores massif is made up of Miocene basalts; Upper Miocene and Lower Pliocene ash tuffs with interstratified trachyandesites, phonolites, and trachyphonolites, Upper Pliocene andesitic or basanitic plateau basalts, and lower Pleistocene basanites, the latter well preserved on valley floors. Thermal springs are still active in the Montes-Dores area indicating that the Puy de Sansy is the latest of the four major volcanic centers in the massif. The town of Le Mont-Dore is visible in the valley to the north. Photo: J. Green.

PLATE 77B

Basaltic feeder dike in fissure of the King's Bowl Rift, eastern Snake River Plain, Idaho (Plate 83A). Feeder dikes of this type provide the source of lava flows which emanate from the main rift. The presence of symmetrically paired basaltic units in the dike attest to repeated outpouring and drainback of lava into the source vents. Caption M. Prinz. Photo: J. Papajakis

PLATE 78A

Dike swarm in Koolau Volcano, windward (northeast-facing) slope of Oahu Island, Hawaii. Dikes are nested together and very little country rock remains. Caption and Photo: R. S. Fiske, U.S. Geol. Surv.

PLATE 78B

Dike-sill relationships, Sgeir Lang, NW Trotternish, Isle of Skye, Scotland. Flat-lying Jurassic sedimentary rocks, which support both hammer and scale in this photograph, were cross-cut by the dike at the right and intruded by the sill seen at the top of the photograph. This sill and its feeder dike were later separated by another thin sill (behind the hammer) which was intruded along the lower contact of the earlier sill. Caption and Photo: T. Simkin.

PLATE 79A

The Great Whin Sill, seacoast at Castle Point, Embleton Bay, Northumberland, north England. Sills spread to a distance depending on the hydrostatic force with which they are injected, their viscosity and temperature, and the weight of the overlying strata in which they are injected. The quartz dolerite of the Great Whin Sill covers an area of at least 3,890 square kilometers and is intruded in a 60-meter thick unit along the bedding plane of low dipping sandstones (this photograph) and Carboniferous carbonates. The source of the quartz dolerite is a series of east-west trending dikes which have risen along faults. Dike to sill transitions have been observed. Emplacement occurred prior to the Permian since the Great Whin Sill cuts across a Permian conglomerate. Along the surface strike of the Sill, a conspicuous escarpment is produced because of the relative resistance of the Sill to erosion. Crown copyright Geol. Surv. Photo: Reproduced by permission of the Comptroller, H. M. Stationery Office.

PLATE 79B

Exposure of the Triassic Palisades sill along New Jersey Highway 5, 0.8 kilometer north of the Palisades Amusement Park. In places, this sill is almost 300 meters thick. The dark band (4 to 6 meters thick) in the center is the olivine-rich gravity-differentiated layer about 15 meters above the base of the sill. This zone (25% olivine) is described as "rotten" because of its greater susceptibility to chemical weathering. The band overlies the fine-crystalline basalt of the lower chilled zone. Above the band is normal coarser-crystalline basalt (here grouted with a cement coating to prevent rock falls). Photo: C. J. Schuberth, Geology of New York City; and Environs; Natural History Press. Copyright 1968.

PLATE 80A

Sheets of Jurassic tholeiitic diabase in upper Paleozoic sandstone. The lower sill is cut obliquely by an inclined sheet which turns into a concordant sill at top of mountain. Carbonaceous sediments strip the sandstone between the diabase sheets. Note the banding in the lower sill. The cliff is 600 meters high. Finger Mountain, South Victoria Land, Antarctica. Caption and Photo: W. Hamilton, U.S. Geol. Surv.

PLATE 80B

Outer cone sheet of quartz dolerite, center 2 of Arnamurchan, Scotland. Exposure near Kilchoan Bay, South shore of Arnamurchan. The cone sheets are sills intruded into lower and upper Lias shales and sandstones. Cone sheets were injected by upward pressure of a monzonitic magma at center 2. The sheets are cut by radially trending feldspar-porphyric and mafic dikes. Some of the cone sheets are over 10 meters thick; they dip inward toward center 2 at angles between 10 to 30 degrees. The area shown is the subject of Excursion 8 of the Field Excursion Guide to the Tertiary Volcanic Rocks of Arnamurchan (Deer, 1969, p. 28–29). Crown copyright Geol. Surv. Photo: Reproduced by permission of the Comptroller, H. M. Stationery Office.

PLATE 81A

Lower southeast slope of Silali Volcano, Baringo District, Kenya (Gregory Rift Valley), showing a very youthful (uppermost Pleistocene to Holocene) closely-spaced grid fault pattern. The normal faults traverse alkalic olivine basalt flows and toward the west have thrown them down about 60 meters, forming near vertical escarpments. Trains of small post-rift blowhole craters follow the fault traces (left and right center; north is to the right). Aerial photo taken from altitude of about 6,700 meters and encompasses an area 0.9 kilometer on a side. Caption; G. J. H. McCall; Photo: Hunting Aero Surveys (Flight 15 KE 7 018, Dec. 1955) Directorate of Overseas Surveys, Tolworth, England.

PLATE 81B

Open crack, part of the Southwest Rift Zone of Kilauea, Island of Hawaii. This fracture system extends for over 25 kilometers from the summit of Kilauea. In places, individual rifts may be over 15 meters wide and is visibly open to depths exceeding 20 meters. The rift segment in the photo cuts through bedded ash deposits from the rare explosive eruptions of Halemaumau. Caption and Photo: R. Decker.

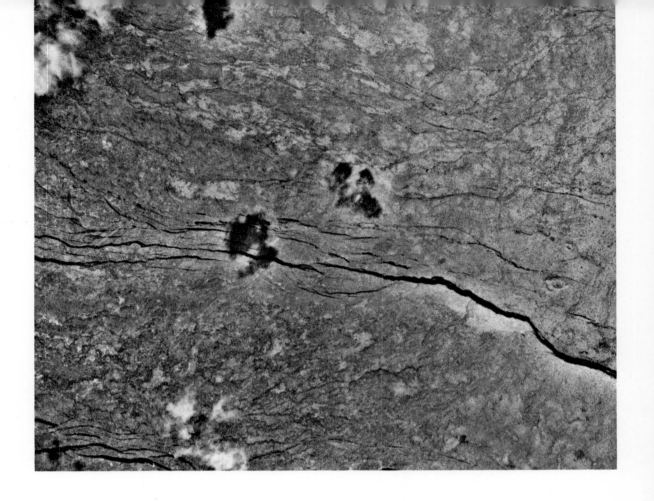

PLATE 82A

The volcanic plateau of the Afar depression, a region in southeastern Ethiopia, west of the Straits of Babel-Mandeb and north of the Gulf of Tadjoura (approximately 12° N and 42.5° E), at the west end of the Gulf of Aden, as seen in space from Apollo 9 (March 11, 1969). North lies close to the bisector of the upper left corner of the photo. The plateau consists of late Tertiary and Quaternary basalts and rhyolites, of cumulative thickness of 1,200 meters and more. The Afar is a large graben structure in which major fractures are related to subsidence following the fissure eruptions that covered the underlying sedimentary sequence. Many of the northeast and north fractures visible in the photo are actually step faults. Note the zig-zag pattern of some of the drainage channels. The base of the photo covers a linear distance of about 190 kilometers. Caption: P. Lowman. Photo: NASA.

PLATE 82B

Aerial view looking southwestward across part of the Koae fault system, Kilauea Volcano, Hawaii. Dilation of the volcanic edifice has produced this zone of grabens, normal faults, and monoclines; very little lava has been erupted from associated fissures. Caption and Photo: R. S. Fiske, U.S. Geol. Surv.

PLATE 83A

The King's Bowl Rift, eastern Snake River Plain, Idaho. This fracture, in places about 2 meters wide, and open to depths up to 280 meters, is part of the Idaho rift system that strikes north to northwest along a total length of over 90 kilometers. Its best known segment extends through Craters of the Moon National Monument. The King's Bowl Rift Set, extending about 8.5 kilometers, consists of central fissures from which volcanic products have erupted, flanked on both sides by subsidiary open fissures without any volcanic material. Successive surges of basaltic lava, followed by subsequent back-draining, result in pairs of lava coatings on opposite sides of the rift. Feeder dikes, now filled with lava surge products, can be seen within parts of the rift (Plate 77B). The most prominent feature on the central zone is the King's Bowl, a phreatic explosion crater, 80×32 meters wide, with vertical walls. The light colored patch east of King's Bowl is ash blown eastward by prevailing winds. Other features on the rift are vents, spatter cones, and ramparts. Caption: M. Prinz. Photo: U.S. Dept. of Agriculture.

PLATE 83B

Part of a row of explosion craters on the west flank of Mount Etna, in Sicily, produced during the eruption of 1910. The three craters shown here share common rim walls. Several other smaller craters follow the trend along which the group developed. A lava flow begins at one end of the crater row, suggesting that the craters are aligned along a fissure from which escaping gases excavated these holes. Photo: Geophys. Lab., Carnegie Inst., Washington, D.C.

PLATE 84A

Fissure eruptions from the Southwest Rift Zone on Mauna Loa volcanic complex, Hawaii (aerial photo at 1:41,200). Hawaiian eruptions of basaltic lava begin with emission of gas-charged lava from long fissures or rifts. Several undated rift eruptions and parts of several historic eruptions can be traced in the photo to the main southwest rift zone (passing from center base of photo to upper right corner) or to parallel, subsidiary rifts on either side. Eruptions in 1950 along a rift (to the left of the main zone) produced the darker, westward flows. Several cinder cones and lines of spatter cones (sites of fire fountains) lie along the main rift; spatter ramparts and separated spatter cones develop along some of the subsidiary rifts as well. Aligned craters (depressions) may be collapsed lava tubes. The 1950 eruption which began on June 1 produced 495 million cubic meters of lava which covered 90.4 square kilometers. Since the vent is only 2,440 meters above sea level, the area covered included a sizable portion of arable land. Photo: U.S. Geol. Surv. Prof. Paper 590.

PLATE 84B

Effect of faulting along the East African Rift Zone in Kenya on the localization and shapes of associated volcanic structures. Vents of the scoria cones and basaltic tuff-rings of the Elmenteita Cones are aligned along pre-existing fracture zones. Intersections of lateral sector grabens, controlled by regional fault trends, form the centers over which the rings, cones, and fissure flows tend to develop. The volcanic craters are elongated in the same directions as the prevailing fault line trends. Caption: G. J. H. McCall. Photo: 82nd Squadron Royal Air Force.

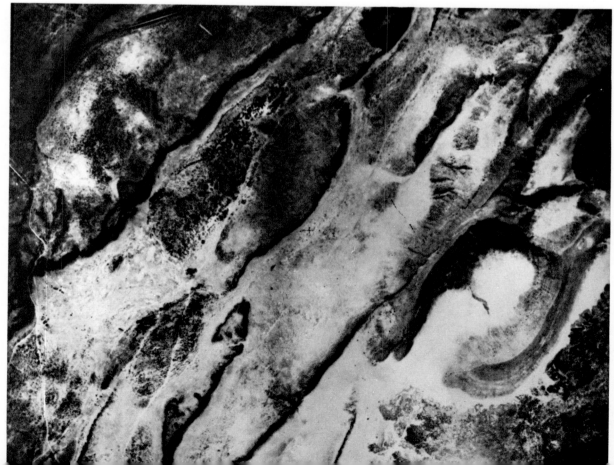

PLATE 85A

Northeast section of Lakagigar crater row, southcentral Iceland, view to southwest. This fissure eruption along the Mid-Atlantic Ridge fracture zone has produced a 25-kilometer alignment of some 100 spatter cones and conelets of tholeiitic basalt, many mantled with ash and lapilli expelled from the 1783 eruption which produced the alignment. During the summer of 1783 about 12 cubic kilometers of basalt were extruded covering an area of 565 square kilometers. The volume of tephra produced was about an order of magnitude smaller. Photo: S. Thorarinsson.

PLATE 85B

Alignment of maar craters, south central Iceland, northeast of Landmannalaugar (Veidivätn area). Litlisjor Lake (center of photo) is 7.5 kilometers long. Craters lie along tension fractures corresponding to the surface expression of the fractured crest of the mid-Atlantic ridge. Photo: J. Green.

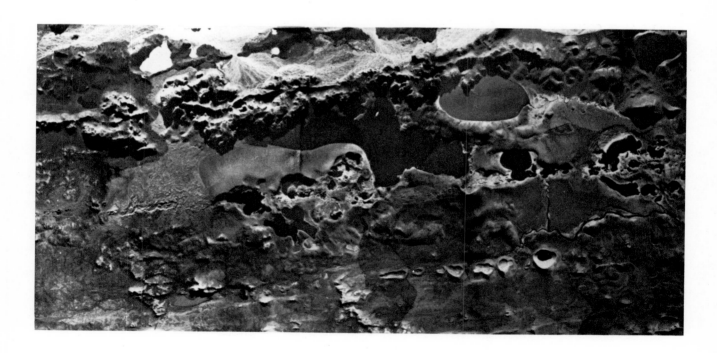

PLATE 86A

Oblique aerial view of Mount Tarawera, North Island, New Zealand, elevation 1,150 meters above sea level. The series of explosion pits in the foreground and center formed in six hours in the evening of June 10, 1886 when the three domes of Tarawera were split by the eruption. The eruption extended to Rotamahana destroying the famed pink and white terraces. Lake Rotomahana itself was drained and the lake bed level was lowered from 10 to 70 meters. Some of the bombs consist of cores of rhyolite thinly coated with basalt. Rhyolitic ash underlies the basaltic ash in the rift. The mountain had been dormant for 900 years. The blast, heard in Auckland, killed over 150 people. Photo: S. N. Beatus, New Zealand Geol. Surv.

PLATE 86B

Tarawara rift showing alignment of explosion pits formed in 1883 and mantling of rift walls with basaltic ash overlying rhyolitic ash. Note the absence of any translational movement of the rift walls. This can be verified by walking on several "bridges" of pyroclastics which connect one side of the rift with the other. Photo: D. L. Homer, New Zealand Geol. Surv.

PLATE 87A

Spatter cone alignment oriented northwest localizing a breached cinder cone, Craters of the Moon National Monument, Idaho. The cinder cone is 25 meters high. Photo: J. S. Shelton.

PLATE 87B

Alignment of domes of rhyolite obsidian, South Sisters Quadrangle, near Bend, Oregon. Domes are aligned on a north-south trending fracture that passes to the east of Devil's Hill (middle of photo) into South Sister. The domes are on older glaciated andesite and rhyolite lavas. The northernmost dome on this six-kilometer alignment is the source of rhyolite flow which extends to older (Pliocene) olivine basalt and basaltic andesite to the east. In the background, South Sister, a composite basaltic stratocone, is seen (elevation 2,440 meters). Two basaltic cones south of the Cascade Lakes highway (in foreground), also of the same approximate age as the rhyolite domes, are aligned parallel to the dome axis but displaced to the west about a half a kilometer. Photo: Oregon Dept. Geol. Mineral Ind.

PLATE 88A

The contact between two flow-units of the Suwanee basalt flow (Quaternary), Valencia County, New Mexico. As defined by Nichols, a flow-unit is a specific outpouring of lava, a few centimeters to several meters thick and of variable lateral extent, that forms a subdivision of a single flow; a series of separate but penecontemporaneous overlapping flow-units, each more or less tongue-shaped, together comprise a single flow. In the photo shown, the lower unit is vesicular and jointed near its top whereas the upper unit has only a few vesicles in its chilled margin where it fitted into the irregular surface of the older unit. Photo: R. L. Nichols.

PLATE 88B

Contact of two basalt flows, north side of Yellowstone Highway out of Bear Gulch, Fremont county, Idaho. Hammer straddles contact. The difference in the geometry of the vesicles in the lower part of the upper unit and the upper part of the lower unit may be due to variation in the volatile content of the lava. In the case of the lower flow, the lesser pressure near the flow top could have caused gases to come out of solution producing more or less spherical vesicles. In the case of the upper younger flow, the surface on which it moved may have been unusually moist causing gases to evolve at the base of the flow, probably shortly after emplacement but before solidification. The gases were able to move upward to form basal pipe vesicles (see Plate 162 D). Photo: H. T. Stearns, U.S. Geol. Surv.

PLATE 89A

Tholeiitic basalt lava blocking Chain of Craters Road, Island of Hawaii. The low, spread out tongues of the flow front are characteristic of the more fluid, basic lavas. The lava extruded from a fissure extending from the northwest rim of Alae Crater to near the base of Kane o Hame, producing two flows which rapidly moved south to cover the Chain of Craters Road. Flows began on February 22, 1969. On that date, the flow was preceded (at 06:27) by a swarm of short tremor bursts and small earthquakes; the lava appeared at 09:50. This lava is vitreous and is covered with a hackly skin. Photo: J. Green.

PLATE 89B

Front of a slow moving and crumbling blocky lava flow at Parícutin Volcano, 1945. The steepness of this front is typical of the more silicic, viscous lavas. Dust arises as the loose material cascades down the front of the flow. Eroded ash beds appear in the foreground. Caption and Photo: T. Nichols.

PLATE 90A

Northwest slope of Silali Caldera, Baringo District, Kenya (Gregory Rift Valley). Arcuate scarps of phonolitic trachyte flows of Pleistocene age cover most of the area shown. Older tuff rings and trachyte lava cones are overridden by the trachyte flows (lower right). The scalloped wall, floor, and internal basalt scoria cones of Silali Caldera appear at the upper right. Cascades of alkalic olivine basalt, later than the phonolitic trachyte flows, are seen forming northward (north to the left) flowing tongues at the upper left. Aerial photo from 6,770 meters altitude; land surface approximately 1.25 kilometers on a side. Caption: G. J. H. McCall. Photo: Hunting Aero Surveys (Flight 15 KE 9, Dec. 1955) for Directorate of Overseas Surveys, Tolworth, England.

PLATE 90B

Big Obsidian Flow, Newberry Caldera, Oregon. The flow emanates from a dome localized on an east-west fracture bounding the southern side of the Newberry Caldera. The corrugations (corda) shown on the flow surface are about 5 meters in wave length and some 2 meters in amplitude. Their morphology is related to flow viscosity and flow rate. The angle of the flow terminus is from 40 to 80 degrees which is far greater than that of the lower viscosity basaltic flows in the area. A detailed account of the chemistry and petrology of the flow is given by Green (1965). Photo: Oregon Dept. Geol. Mineral Ind.

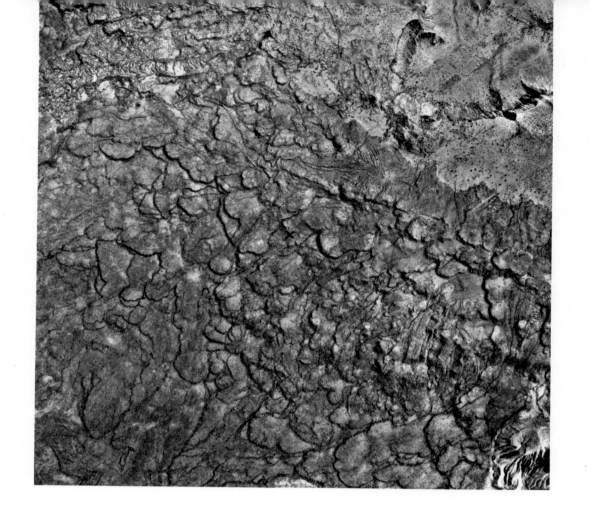

PLATE 91A

Pakka Volcano, Kenya. A trachyte flow with terminal cliffs and concentric pressure ridges is seen to emanate from the throat of a breached cone. Caption: G. J. H. McCall. Photo: Hunting Aero Surveys for Directorate of Overseas Surveys, Tolsworth, England.

PLATE 91B

Closer view (1:20,000) of the viscous silicic pantelleritic pitchstone flow unit erupted down the northeastern flank of Fantale volcano (see Plate 22A). The main directions of movement are emphasized by the curved ridges that result from differential velocities of flow. Caption and Photo: I. Gibson. Imperial Highway Authority of Ethiopia.

PLATE 92A

A lava field emanating from several irregularly shaped crater vents at Pillar Butte, Idaho. An extensive network of lava channels and dikes carry the basaltic flows to considerable distances beyond the source. The photo width is 3.6 kilometers. Caption and Photo: R. Greeley.

PLATE 92B

Rough aa lava topography on the western portion of the Pinacate Mountains, Sonora, Mexico. In the upper western portions of this field, cinder cones predominate; in the lower portions, maar craters. Photo: T. Nichols.

PLATE 93A

Recent basaltic flows, associated with pyroclastic cones, western of the Snaefellsnes Peninsula, western Iceland. Note steepness and smoothness of the cone slopes and the large, coarse corda developed on the flow surfaces. Photo: Landmaelingar Islands contributed by J. Green.

PLATE 93B

Lobate ends of basaltic lava flows from the Hekla Volcano in Iceland. These aa flows are about 6 meters thick and consist of a basal breccia, a massive core, and a breccia flow top. Caption and Photo: R. Decker.

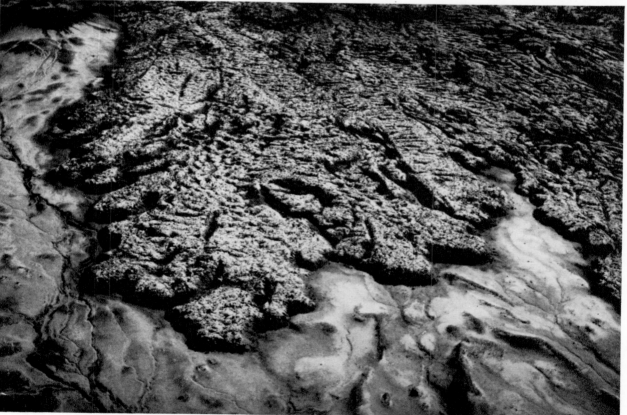

PLATE 94A

Part of the Craters of the Moon National Monument, Idaho, as seen in an aerial photograph. Basaltic cinder cones, many of which are aligned along a northwest-trending rift zone, have conspicuous craters of various sizes (see lower half of photo). In places, the rift zone is marked by a line of small spatter cones (Plate 87A). The craters are surrounded by dark basalts (center left and right), issuing from the rift zone. Older basalt flows, in part covered by vegetation, appear in the north (upper) half of the photograph). Photo width covers 4.3 kilometers. Caption and Photo: U.S. Geol. Surv. Prof. Paper 590.

PLATE 94B

Aerial view of northeast rift of Mauna Loa volcanic complex, Hawaii. Young pahoehoe lava from vents upslope (upper left) is deflected by a spatter cone of late prehistoric age, producing a distinct wake effect in the flow pattern. Caption and Photo: R. S. Fiske, U.S. Geol. Surv.

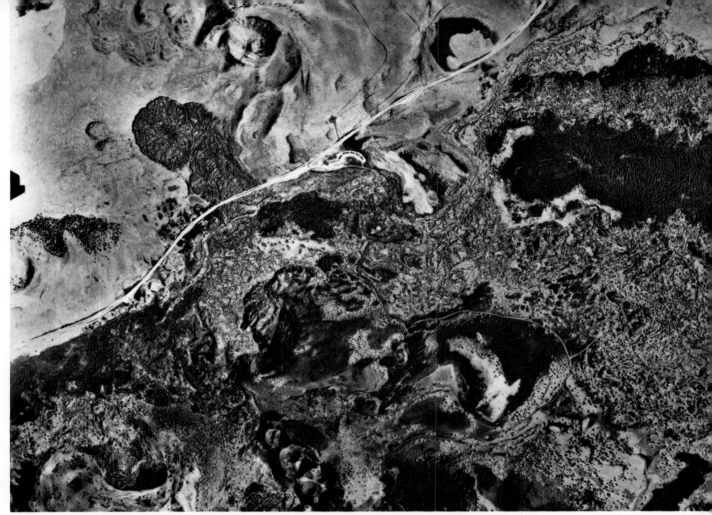

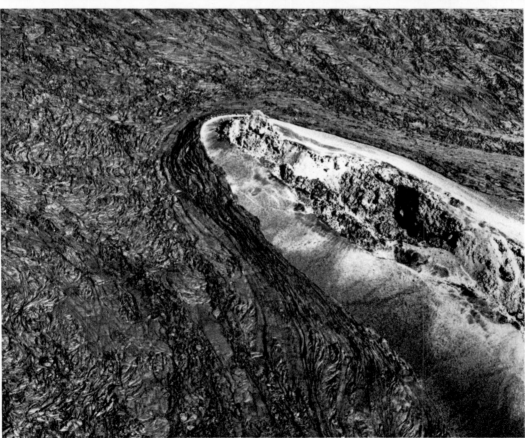

PLATE 95A

Kipukas or steptoes, patches of older terrain, in this case a reddish lava surface, surrounded by more recent, darker basalt flows which built up a small shield volcano, view to northwest from McKenzie Pass, Highway 26, northwest of Bend, Oregon. The island-like steptoes are remnants of basaltic cinder cones developed around the double-cratered cinder cone, Belknap Crater (in right background), erected by pyroclastic eruptions at the summit of the shield. The younger engulfing flows, although of pahoehoe character, were broken into blocks and slabs (foreground) that piled up in a hummocky, rough surface; this disruption resulted from the steep gradient of the flow together with crustal foundering as lava drained out from below. Photo: Oregon State Highway Travel Division.

PLATE 95B

Northwest Rift Zone, Newberry Caldera, Bend Oregon. A series of faults and fractures localizing basaltic lava trending North 30 West. At least eight separate basaltic flows have erupted from this zone. One of these is the Lava Cast Forest Flow dated by charcoal inclusions in the flow to be $6,150\pm210$ years old (N. Peterson and E. Groh, unpublished data). Lava Cascade Flow is probably younger and originated by fire-fountaining as the morphology of spatter cones and ramparts suggests. Photo: Oregon Dept. Geol. Mineral Ind.

PLATE 96A

Flow units of basalt, Four Falls, Yellowstone National Park. The intercalation of these flows with continental gravels attests to the rapid uplift of the area during the period of basalt extrusion. Many of the flows show columnar jointing. Aerial erosion followed the outpouring of basalt; the flows shown are being exposed by the downcutting action of the Yellowstone River. Photo: G. A. Grant, Nat. Park Serv., U.S. Dept. Interior.

PLATE 96B

Caprock of jointed basalt, one of a series of thin flows (see background) within clastic sediments (lake beds, alluvial plains, etc.), between Tucannon River and Riparia, Washington. Such intracanyon flow remnants can exert strong influence on local and regional topography. Photo: F. O. Jones, U.S. Geol. Surv.

PLATE 97A

Columnar basalt flows, Grand Coulee, southeast Washington. Because each of the flow units shown on the east wall of this coulee came from different vents at different distances, the viscosity of the basalt varied to the degree that the development of the columns also varied. Fairly long time intervals separating the flows is indicated as proven by the fossil soil layers or tree stumps found on the upper surfaces of unit flows. The figure is instructive in that it shows the average thickness of a flow unit of tholeiitic basalt does not exceed 10 meters; the exposure is 160 meters high. Contacts between flows vary also, and these may represent local erosion of the lower flow prior to the later one or wedging out of flow units. That the orientation of columns changes from place to place may be a function of variations in cooling rates which in turn may reflect different concentrations of volatiles in the original lava and the surface traversed. The greatest thickness of these late Tertiary tho'eiitic Columbia River Basalts is in the Snake River Gorge where 1,200 meters of lava are exposed. Photo: J. S. Shelton, from *Geology Illustrated* by J. S. Shelton, W. H. Freeman and Company, Copyright 1966.

PLATE 97B

Basalt flows in the Columbia Plateau, along the Snake River, downstream from Palouse Falls, Franklin and Whitman counties, Washington. Note the effect of jointing in controlling stream course. Hill in right background is Palouse loess rising above basalt surface. Caption: F. O. Jones. Photo: U.S. Geol. Surv.

PLATE 98A

Lahar mounds along the Chateau Road on the northwest side of Ruapehu Volcano, Central Volcanic district, on North Island, New Zealand. Lahars are mudflows that result from the mixing of ash, rock debris or nuées ardentes deposits with river water, runoff from heavy rains along the slopes of volcanoes, or sudden breaking of crater lake rims allowing rapid release of large volumes of water. Lahar deposits are poorly sorted, chaotic masses of debris, ranging from boulders (meters in diameter) to clay, that form volcanic conglomerates or breccias (see Plate 174C). The mounds develop by compaction and draining of the lahar deposits; commonly, boulders underlie the hummocks. Photo: R. S. Dibble.

PLATE 98B

Mudflow levees on the northwest slope of Shastina, the cone just west of Mount Shasta, shown in an aerial photograph. These levees extend from the bare debris-covered slope down into the forest. Some levees are forested in their lower courses and are ancient; other are bare and mark channels down which debris of volcanic character has moved in recent years. Caption and Photo: U.S. Geol. Surv. Prof. Paper 590.

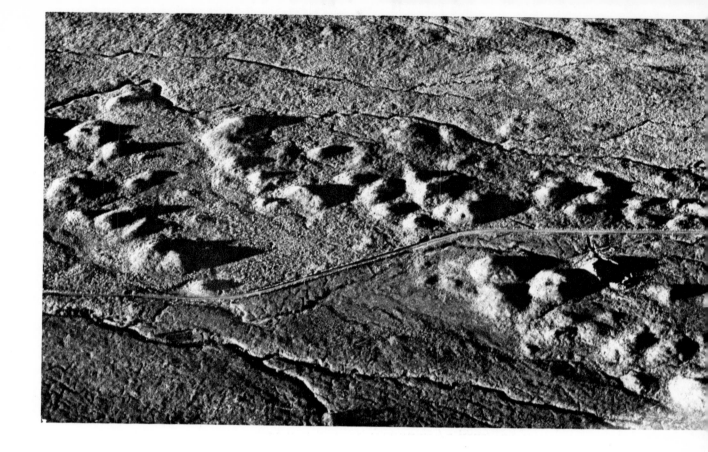

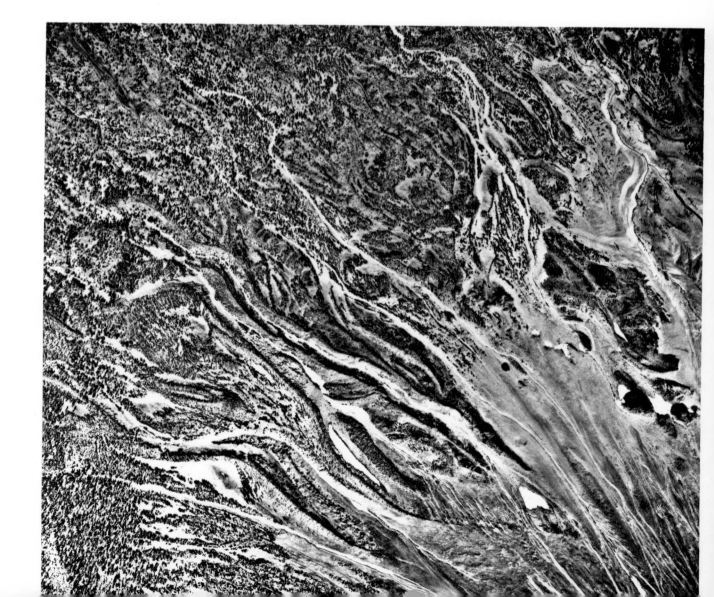

PLATE 99A

Lobate terminus of a small mudflow along outer flanks of Vesuvius, near Naples, Italy. The furrows result from later dissection by rainwater. Caption and Photo: Geophys. Lab., Carnegie Inst., Washington, D.C.

PLATE 99B

Lahar mounds of hydrothermally altered andesitic rubble carried down into the breached portion of White Island Volcano, Bay of Plenty, New Zealand. (see Plate 33B). Blocks of porphyritic andesite, in which the feldspar phenocysts are partly kaolinized, are up to 1 meter in diameter. Some are veneered with sulfur-rich gypsiferous clay. Photo: F. O'Leary, Dominion Observatory, New Zealand.

PLATE 100A

Recently fallen tephra (ash) from the Syrtlingur island (the second island formed during the Surtsey eruption), covering a pahoehoe lava field on Surtsey Island. (July 24 1965). Caption and photo: S. Thorarinsson.

PLATE 100B

Basaltic ash, Parícutin, Mexico. View to north in gulch southwest of Parícutin cone. Exposure is 3.5 meters high. The high angle of repose of the layered basaltic ash is due partly to the moisture content of the ash and partly to its angular grain shape. The material shown was used in simulated lunar soil research (Osgood and Green, 1966). Note that the material within the layers is very well sorted because of the ejection of ash within a small size range for sustained periods of eruption. Photo: J. Green.

PLATE 101A

Aerial view looking west-northwest into the Valley of Ten Thousand Smokes, Katmai National Monument area in south-west Alaska (position of viewer near Mount Katmai Volcano). The valley (right and upper right), now occupied by up to 300 meters of tuff deposited during a single explosive event on June 6, 1912, is approximately 22 kilometers in length and 3 to 5 kilometers in width. The valley is bordered with mountains composed of Jurassic sedimentary rocks and later volcanics; the Knife River drains the present valley which had been widened by previous glaciation. The tuff units also extend laterally to the southwest and northeast in the broad area left and right of the hills in the center of the photo. The source of most of the tuff is the small Novarupta crater with its central cumulo-dome or tholoid of rhyodacite (see Plate 68B), visible in the upper left center. Falling Mountain and Mount Cerebus appear just beyond (left). Novarupta and Broken Mountain (front) and Baked Mountain (back) lie to the right. In the foreground of the photo are several active glaciers that are moving downslope from Trident and Mount Katmai Volcanoes. Photo: U.S. Air Force.

PLATE 101B

Part of the great pyroclastic deposit composed of rock fragments and ash that fills the Valley of Ten Thousand Smokes, west of Mount Katmai, Alaska. This ash flow or nuée ardente at Novarupta (see Plate 101A) formed an active vent during the June 6–7 explosive eruptions at Katmai. The tuff flow was emplaced on June 6. Major explosions at Katmai on June 6 and 7, 1912 enlarged a caldera which eventually was occupied by a crater lake. At least five ash layers in the Valley can be traced to explosive eruptions during these three days; four of these are traceable to Novarupta. The tuff flow filling the Valley and surroundings has a volume estimated by G. H. Curtis to be about 11 cubic kilometers. View in the photo is to the west, with Knife Creek in the foreground. Photo: U. Clanton, NASA, Manned Spacecraft Center, Houston.

PLATE 102A

The edge of a pyroclastic flow from one of the vents of Sheveluch Volcano in the Kurile Islands. The nuée ardente consisted of ash mixed with rock fragments and blocks of varying sizes. The rock fragments are largely hornblende andesites. Caption and Photo: G.S. Gorshkov.

PLATE 102B

Terminus of ash flow, Kurile Islands. The photograph illustrates the thinness of a rubble and dust flow lubricated by a relatively large volume of gas. The release of the gas phase dropped the suspended load leaving a veneer of pyroclastic blocks and boulders. This is in contrast to other volcanic flow types where a liquid phase carries the material, resulting in flow termini much thicker than that shown here. Photo: G.S. Gorshkov.

PLATE 103A

Volcano Island in Lake Taal (which fills an old caldera) in southwest Luzon, Philippine Islands. In 1965 a series of phreatic explosions produced the elongate crater (1.7 by 0.8 kilometer from which a partly submerged cone is emitting steam.). At the height of the eruption, ash and steam clouds moved high into the atmosphere. During violent explosions debris-laden clouds also spread laterally in all directions at high velocities, acting much like the base surge that accompanies a nuclear cratering explosion. (See Plate 3B). Over a radius of 4 kilometers, ash dunes (see foreground) were built up by deposition of coarser ejecta. Photo: J. G. Moore, U.S. Geol. Surv.

PLATE 103B

Aerial photograph toward the east showing new explosion crater of Taal Volcano (extreme right). Ash dunes concentric to the explosion crater are visible in the foreground. Thus, these dunes are oriented roughly at right angles to the direction of movement of the base surges. Dune crests in each overlying bed are slightly displaced away from the explosion center as the deposit builds up by deposition from successive base surges. The dunes are largest adjacent to the explosion crater where they attain wave lengths up to 19 meters. Wave length systematically decreases away from the crater to about 4 meters at a 2.5 kilometers distance. Caption: J. G. Moore. U.S. Geol. Surv. Photo: R. Moxham, May 1966. U.S. Geol. Surv.

PLATE 104A

A small, crusted lava lake, slightly depressed below its rim, in the floor of Halemaumau pit crater at Kilauea, Island of Hawaii on Sept. 7, 1920. Other active areas, where the floor has foundered, appear to the left. Photo: Geophys. Lab., Carnegie Inst., Washington, D.C.

PLATE 104B

Aerial view of lava lake in Makaopuhi Crater, Kilauea Volcano, Hawaii (Plate 44B). The lava lake has a maximum depth of 80 meters and was formed during the eruption of March 1965. Note the spatter cone on the lower left. Shadowed cliff in upper part of photograph is section through a prehistoric lava lake that partly filled an ancestral Makaopuhi Crater. Makaopuhi Crater is 1.4 kilometers wide in an east-west direction. Caption and Photo: R.S. Fiske, U.S. Geol. Surv.

PLATE 105A

Vertical aerial view of Aloi Crater, Kilauea Volcano, Hawaii, shortly after the eruption of December, 1965. Lava erupted from fissures that cut the crater (steaming areas) and the floor of the crater was buried to a depth of about 1.5 meters. The crater is 175 meters in diameter. Caption and Photo: R. S. Fiske, U.S. Geol. Surv.

PLATE 105B

Aerial view of the 1877 lava lake in Keanakakoi Crater just outside southeast edge of Kilauea caldera, Island of Hawaii. Troughs on the surface are filled with light-colored scoria that emphasizes the outlines of polygonal hummocks bounded by these troughs. The ash-covered lake is approximately 230 meters wide and 385 meters long. Caption: D. Peck. Photo: Elliot Endo, U.S. Geol. Surv.

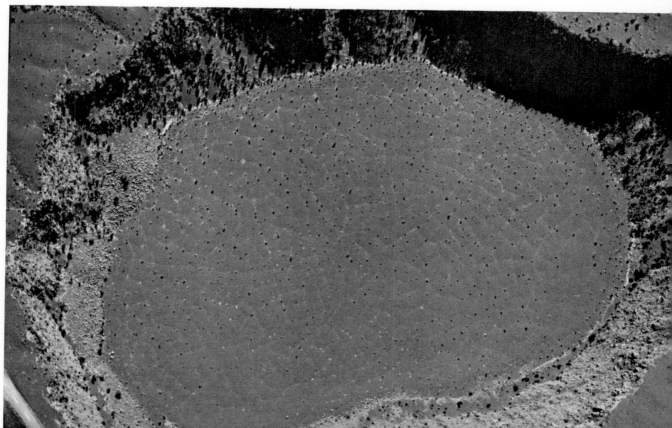

PLATE 106A

Cascading lava in lava river in floor of Halemaumau pitcrater, Kilauea caldera, Hawaii. Flow rate is about that of light oil (i.e. viscosity of about 300 poises). The glowing streaks are the result of time-lapse photography. A drainback phase of lava into a rift on the craterpit floor was in progress when the photo was taken. Photo: Geophys. Lab., Carnegie Inst., Washington, D.C.

PLATE 106B

A "frozen" lava river in a field of pahoehoe lava near Kilauea, Island of Hawaii. A stream of lava, possibly draining from the tumulus (mound in center), formed a shallow channel on the surface. Its top now shows filamented pahoehoe and spatter deposits formed during the flow. Photo: Geophys. Lab., Carnegie Inst., Washington, D.C.

PLATE 107A

Confluence of two lava streams, Craters of the Moon National Monument, Idaho. The flow lines of the frozen pahoehoe lava indicate the direction of flow with the cross corrugations convex downstream. Ridges built up parallel to the stream edges indicate fluctuations in the level of the lava in the stream. Photo: R. Menning, Nat. Park Serv., U.S. Dept. Interior.

PLATE 107B

Smooth and ropy helluhraun crust over a 3-meter wide lava rivulet on the floor of Surtur I, Surtsey, Iceland. This olivine basalt-filled channel emerged from a lava tube below spatter cones aligned along the eruption fissure of August 19, 1966. Lava spatter agglutinate and scoria crust are characteristic of the flow surface adjacent to the rivulet. Caption and Photo: J. D. Friedman, U.S. Geol. Surv.

PLATE 108A

Lava filling flow channel, Isabela Island, Galapagos Islands. Lava appears to be draining down and inward into beginning of a lava tunnel. Width of channel is 2 meters and visible length about 10 meters. Corrugation of pahoehoe lava in channel is convex in direction of flow, a feature often seen in lava on the floors of lava tubes. Photo: J. Green.

PLATE 108B

Sheets of high viscosity lava being extruded from near the main lava source, Parícutin Volcano, Mexico (1945). Wrinkles and grooves produce distinct flow markings. Caption and Photo: T. Nichols.

PLATE 109A

A cascade of pahoehoe lava over an inclined surface at the Craters of the Moon National Monument, Idaho. Lines of flow in the now solidified lava are indicated by the ridges and grooves parallel to the extension of the cascade. In the foreground, the coils of ropy pahoehoe are distributed essentially normal to the direction of cascade movement. This ropy structure built up as the lava moved out onto a flatter surface. Photo: R. Menning, Nat. Park Serv., U.S. Dept. Interior.

PLATE 109B

Pahoehoe lava cascades down the slope of a small satellitic cone on the southwest rift of Kilauea Volcano, Hawaii. Parts of the cascade have collapsed, revealing small lava tubes beneath. Caption and Photo: R. S. Fiske, U.S. Geol. Surv.

PLATE 110A

Cascade of pahoehoe lava, Halemaumau, Kilauea, Hawaii. The overflow resulted from filling of a lava lake within the Halemaumau firepit into a lower nonmolten portion of the floor of the firepit. Photo: Geophys. Lab., Carnegie Inst., Washington, D.C.

PLATE 110B

Thin skin of lava erupted in December 1965 encrusting talus blocks at the margin of Aloi Crater, Kilauea Volcano, Hawaii. Caption and Photo: R. S. Fiske, U.S. Geol. Surv.

PLATE 111A

Aerial view along the southwest rift of Mauna Loa, Hawaii. Lava channel in 1950 flow winds out of view to the right. Spatter cones formed during recent prehistoric eruptions cut diagonally across left and center of photo. Caption and Photo: R. S. Fiske, U.S. Geol. Surv.

PLATE 111B

Open lava channels on the west slope of Mount Hualalai, western side of Island of Hawaii. These channels, 10 to 20 meters wide, are partly filled with solidified lava, but they did not overflow or bridge to form tunnels within the field of view. Their sinuosity is characteristic of lava rivers. Scars of older, largely filled channels are evident. Photo: U.S. Air Force.

PLATE 112A

Sinuous lava flow channels, Fire Mountain area, Lanzarote, Canary Islands. The channel, up to 22 meters wide, exhibits undercut walls, possibly of a pre-existing lava tube which has collapsed. Channel floor lies from 5 to 7 meters below its rim. Photo: J. Green.

PLATE 112B

Lava gutters or channels, produced during the 1931 eruption on Tristan da Cunha, a volcanic island on the south Atlantic Ridge. The trench, here about 6 meters wide and 3 meters deep, is a constructional feature built up by overflow of lava. Lava levees are formed by the overflow onto the flanks of the channel which is then left exposed as the supply of trachyandesitic lava dwindles. Block lavas (not aa) are left behind in both the channel and its flanks, on which spiny projections are also developed. The walls of the gutter are marked by horizontal grooves. Formation of such open channels depends on lava temperature, cooling rate, and viscosity; the lava must be neither too viscous nor too fluid and should flow over a moderate gradient. A detailed account of the morphology of the 1931 eruptive center is given by Baker *et al.* (1964, pp. 545–558). Photo: P. E. Baker.

PLATE 113A

Collier Cone, north flank of North Sister, near Bend, Oregon. The cone, composed of basaltic cinders is uniform and smooth on the north slope; the slope of the cone is 30 degrees. Basaltic lava flows, some with well-developed corda, are constrained to flow within lava levees about 10 meters high built up from earlier flows. The flows breached the west wall of Collier Crater, some reaching a length of 13 kilometers to the northwest. In one case, the flows broke the northwest levee and moved toward the northeast. To the south-west of the crater toward North Sister, lateral moraines deposited by the retreat of the Collier glacier are visible. Photo: Oregon Dept. of Geol. Mineral Ind.

PLATE 113B

Lava channels on north slope of Isla Fernandina, Galapagos Islands. Flows issued from concentric feeder vents near the summit (approximately 1,400 meter elevation) and formed large aa channels down the unusually steep (to 34°) flanks. Spatter cones are visible on the horizon (see also Plate 18B). Caption and Photo: T. Simkin.

PLATE 114A

Small channels of pahoehoe-type lava issuing from a spatter cone on the summit of Bartholomew Island, Galapagos Islands. Caption and Photo: T. Nichols.

PLATE 114B

Small filled and empty lava tubes on side of collapse crater walls. McKinney flow, Snake River Plain, Idaho. Caption and Photo: H. Wilshire, U.S. Geol. Surv.

PLATE 115A

Truncated lava flow, south side of Surtsey, Iceland. The seacliff exposed by wave action shows a lava tube that fed the advancing front of a lava flow now eroded away. The viscosity of the Surtsey lava (about 1,000 poises at 1,150°C) permitted rapid flow of basaltic lava even at low gradients. Photo: S. Thorarinsson.

PLATE 115B

Lava tube in the center of a flow lobe of basalt, Santa Ana Mesa, Jemez Mountains, New Mexico. Note the distinctive pattern of radial and concentric joints around the tube. Caption and Photo: R. Bailey, U.S. Geol. Surv.

PLATE 116A

Interior of lava tube, Lava Beds National Monument, California. The "high lava" line of a lava flow that last flowed through the tunnel is visible. "Lavacicles" spattered from earlier flows that more completely filled the tunnel are evident. Fractures in congealed melt from any one lava flow passing through a lava tube extend only to the next inner lava lining of the previous flow. In other words the fracture networks are only lava skin deep. Photo: Nat. Park Serv., U.S. Dept. Interior.

PLATE 116B

Detail of "shoreline" curbs in basaltic lava tube, looking upstream. Flashlight for scale. Valentine Cave, Lava Beds National Monument, California. Caption and Photo: K. Howard.

PLATE 117A

Collapsed lava tube near Butte Crater, Idaho. From the source crater (left, north), the channel extends southward for over 5 kilometers. Most of the main channel remains open, as the roof more or less completely collapsed inward, but to the south (out of photo), the collapse was incomplete so that disconnected tube segments lie between enclosed cave openings. Flow divergences of the main channel are indicated at the arrow. Over most of its length, the open channel is from 5 to 10 meters deep and 20 to 35 meters wide. No "downstream" displacement of collapsed roof fragments occur, i.e., the open channel formed after all flowage had ceased. Photo: U.S. Dept. of Agriculture, contributed by J. Green.

PLATE 117B

Oblique aerial view of the collapsed lava tube shown in Plate 117A, originating from the Butte Crater, a small (diameter about 100 meters) maar-like crater near the Island Park Caldera, Idaho. Photo: J. W. Dietrich, NASA.

PLATE 118

An aerial photo mosaic (scale at 1:20,000) covering most of the Lava Beds National Monument in Northern California. The nearly circular cinder cone, Schonchin Butte, appears below the photo center. Schonchin Butte Flow (basalt) extends northward for more than 10 kilometers. Near the lower left of the photo, the small, dark ovoid spot is Modoc Crater, from which the Modoc Lava Tube emerges. Collapsed segments of this tube can be traced north, then east past Schonchin Butte, and north again (to the upper right corner) for more than 15 kilometers. Other tube systems (and a fault scarp) are visible on the left side of the photo. Caption: R. Greeley. Photo: U.S. Dept. of Agriculture, contributed by R. Greeley.

PLATE 119A

Mammoth Lava Tube, Lava Beds National Monument, California. Formed in the Pleistocene to Recent Modoc Basalt, Mammoth Crater and lava tube extend from the far northern base of the Medicine Lake Highlands onto the Modoc Plateau. The beginning of the branching distal end of the tube is marked by the two main collapsed branches in the upper right quadrant of the photo. This part, known as the Cave area in the Monument, is highly complex and represents feeder tubes for the advancing lava flow. Caption: R. Greeley. Photo: W. Quaide.

PLATE 119B

Giant Crater and associated lava tube, Siskiyou County, California. The system is on the southern flank of the Medicine Lake Highlands and is the probable source of the Modoc basalt (Pleistocene to Recent) which fills a large north-south graben. Giant Crater is about 235 meters wide, 320 meters long and 52 meters deep. Interior terracing represents subsidence. The initial segment of the tube system probably was roofed only by a thin crust which collapsed during or immediately following the molten stage. A small aa flow extends from the crater rim towards the southeast. A short lava tube (extreme right edge of photo) extending from Chimney Crater, 1.1 kilometers to the northeast, may drain into Giant Crater. Caption and Photo: R. Greeley.

NORTH

CAVE AREA

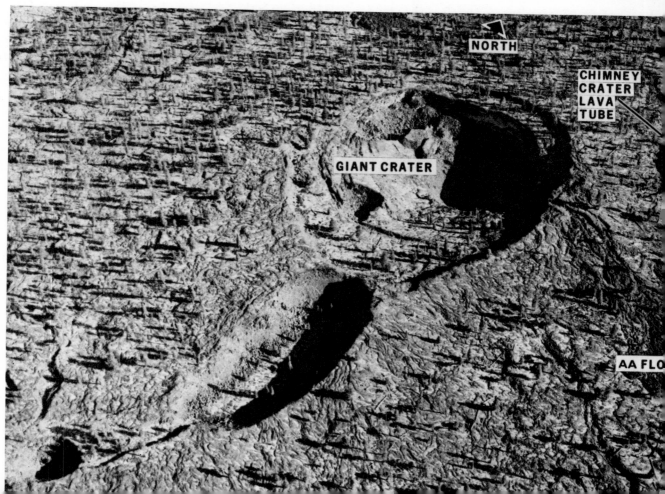

NORTH

CHIMNEY
CRATER
LAVA
TUBE

GIANT CRATER

AA FLO

PLATE 120A

The interior of a large, partly collapsed lava tube (locally called an "Ice Cave") in the Craters of the Moon National Monument, Idaho. Photo: W. Keller, Nat. Park Serv., U.S. Dept. Interior.

PLATE 120B

Surface expression of a series of collapse depressions in the McCarty's Lava Flow, southeast of Grants, New Mexico. The collapse results from drainage of lava chambers, such as the lava tubes which underlie this pahoehoe basalt flow. The blocky fragments filling the depressions represent parts of the roof above a tube. From the air, concentric open fractures surround the depressions. Caption: G. Kuiper. Photo: Univ. of Arizona Lunar and Planetary Laboratory, Tucson, Ariz.

PLATE 121A

Collapse depression on a recent basaltic lava flow, Malheur County, Oregon. As described by N. Peterson, the smooth pahoehoe surface collapsed into pits from 10 to 300 meters in diameter when the subsurface lava drained away. Some of the pits appear to be interconnected or aligned; some are not circular. Photo: Oregon Dept. Geol. Mineral Ind.

PLATE 121B

Craters of the Moon National Monument, Idaho. An irregular pahoehoe surface shows the effects of partial outdraining and shrinkage cooling of the lava. Broad upswells (tumuli) are interspersed with wavy channels (overlying subsurface tubes) and circular to elongate drainage sinks. Fractures were produced on surface layers which bent during collapse. Photo: J. S. Shelton.

PLATE 122A

The Diamond Craters area, Harney County, in southeastern Oregon. Relatively recent volcanic activity has produced a near-circular field of basaltic lavas (10-kilometers diameter) which reaches a thickness of 30+ meters in its center but thins out to a few meters at the edges. The shape of the field is somewhat influenced by irregularities in the underlying topography (older flows and sedimentary rocks). Several large, flat domes (upper left) associated with lobate lava flows formed over shallow intrusions. Drainage through lava tubes developed shrinkage cracks in the pahoehoe crust, followed by collapse into the voids to produce circular to oval sinks and depressions up to 50 to 75 meters long. Photo: Oregon Dept. of Geol. Mineral Ind.

PLATE 122B

Diamond Craters volcanic structures, southeast Oregon. The complexly pitted dome was produced by maar type explosions following the rise of a large domal structure which defines the outer limits of the funnel-shaped pit area. Some 17 probably contemporaneous steam explosions covered much of the surrounding area with basaltic ash and lapilli. Cored bombs described by Peterson (1963) from this area attest to the shallow depth of sediments probably providing the water for the explosivity and material for the basalt bomb sediment cores. In the background (top), Graben Dome is clearly seen. The fracture pattern is the result of longitudinal collapse of a dome downdropping the summit about 30 meters to form 1.5-kilometer long trough. Photo: Oregon Dept. Geol. Mineral Ind.

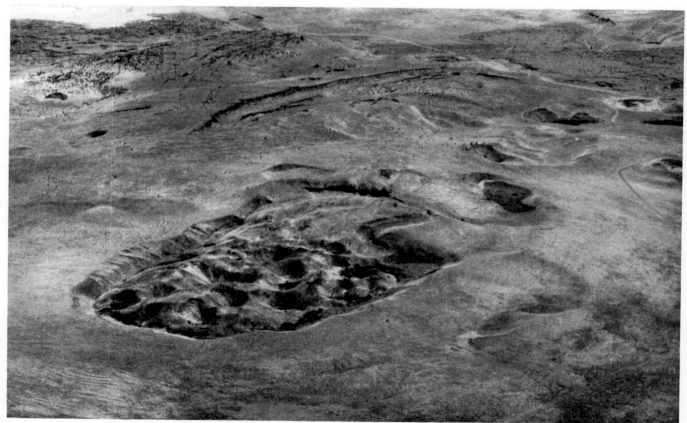

PLATE 123A

Fire fountains of 1925 southwest rift eruption on Mauna Loa, Hawaii. Numerous olivine basalt flows are beginning to develop downhill from the rift; spatter ramparts are also building up around the fountains. The fountaining which began on April 10th, 1926 was from a lower elevation (2,320 meters) than either the preceeding 1919 or succeeding 1933 events. The flank flows which lasted two weeks covered 28.2 square kilometers above sea level. The volume of lava that entered the sea was estimated to be 1.2 million cubic meters. Photo: U.S. Air Force.

PLATE 123B

Building of spatter cones along the East Rift Zone of Kilauea, Island of Hawaii, Volcanoes National Park, near Chain of Craters road leading to Aloi Crater. Photo: D. W. Reeser, Nat. Park Serv., U.S. Dept. Interior.

PLATE 124A

The floor of the lava lake at Kilauea Iki, southeast of Mauna Loa, Island of Hawaii. Cooling fractures are evident over the entire chilled surface of the lake. This surface, formed during the 1959 activity, thickened rapidly so that some months after a hard top developed, it was possible to drill through the crust into the molten lava below for sampling and measurement purposes. The upbulged structure in the top center is actually a spatter-cinder cone produced by ejection of material along the pit crater walls close to lake level. Cinders and ash now mantle the top of this cone but slumping has further altered its originally unusual form resulting from pile-up of the ejecta against a steep wall. Photo: Hawaii Visitors Bureau.

PLATE 124B

Last stage of thermal activity of spatter cones aligned along an (August, 1966) effusive eruption fissure in the floor of Surtur I, Surtsey, Iceland, July 21, 1967. Earlier, during simultaneous lava fountain and cauldron activity in August and September, 1966, the two northernmost cones were breached and later merged to form the larger feature seen here. The small central driblet or spatter cone evolved at a late stage in the effusive eruption. Note the accumulation of spatter agglutinate (near center of photograph) on the near side of the cone. The inner tephra slope of the older rim of Surtur I appears across the photo in the immediate foreground. At the time of its last effusive activity, Surtsey was evolving into a small shield volcano of the Icelandic type (note relatively flat flow surface in background). Caption and Photo: J. D. Friedman, U.S. Geol. Surv.

PLATE 125A

Two spatter cones each some 15 meters high in the basaltic flows of Etna, Sicily. Often spatter cones can be much steeper than one would presume from the viscosity of the feeder lava flows because of the increase in viscosity (by cooling) of the airborne spatter which accumulates as steep-sided agglutinated cones. Photo: Geophys. Lab., Carnegie Inst., Washington, D.C.

PLATE 125B

Two driblet spires formed on the floor of Halemaumau pit crater in Kilauea Caldera, Island of Hawaii. These narrow cones build up from spatter and driblets of basaltic lava emerging from a small central vent. A third cone, on the right, has the form of a small hornito. Photo: Geophys. Lab., Carnegie Inst., Washington, D.C.

PLATE 126A

Hornito, floor of Halemaumau, Kilauea, Hawaii. The cone, about 4 meters in height, is fed by molten lava at depth. There is no permanent root. If sufficient pressure were developed to form a spatter edifice much higher than its width, it would be termed a lava spire. Photo: Geophys. Lab., Carnegie Inst., Washington, D.C.

PLATE 126B

Hornitos covered with basaltic ash in the Atrio del Cavello, the "moat" between the inner crater of Vesuvius and the encircling Monte Somma, Italy. High angle dikes intersect layered pyroclastic deposits and flows making up the inner wall of the Somma. Similar dike swarms are common in many basaltic and andesitic crater walls as at Hakone for example. Photo: Geophys. Lab., Carnegie Inst., Washington, D.C.

PLATE 127A

Open active vent along the southwest rift zone of Halemaumau, Kilauea Caldera, Hawaii. The white interior is above 1,100°C. Strictly speaking this structure is not a hornito if it is located on a feeder vent extending to great depths. Photo: Geophys. Lab., Carnegie Inst., Washington, D.C.

PLATE 127B

Extinct hornito, Lava Beds National Monument, California. A sustained evolution of molten lava from a reservoir in the underlying flow is required to form these rootless structures. The erratic distribution of the lava on this hornito is well shown. Photo: G. A. Grant, Nat .Park Serv., U.S. Dept. Interior.

PLATE 128A

Solidified lava from the December, 1965 eruption after it had poured back into a drainback crack on the upper East Rift of Kilauea Volcano, Hawaii. Caption and Photo: R. S. Fiske, U.S. Geol. Surv.

PLATE 128B

Glassy lava draining back into recent fissure on the northeast rift of Mauna Loa volcanic complex, Hawaii. The light coloration results from incorporation of sublimates. Caption and Photo: R. S. Fiske, U.S. Geol. Surv.

PLATE 129A

A large tumulus (schollendome) formed on the floor of the Atrio del Cavallo between the central Vesuvius cone and Monte Somma, near Naples, Italy. This upswelling or intumescence consists of a more or less solid mound of pahoehoe lava. Many tumuli are believed to result from an arching or bulging upward of viscous lava owing to lateral compression developed during flow movements. In some instances, however, a tumulus may form by upward bending of the lava surface pushed from below by outflowing lava. It is also possible that some tumuli are produced in a manner similar to the usually smaller lava blisters (Plate 134) or bulbous squeeze-ups. Photo: Geophys. Lab., Carnegie Inst., Washington, D.C.

PLATE 129B

Tumulus on the floor of Kilauea Caldera, Hawaii. The longitudinal crest fracture is a characteristic feature of tumuli. The other fractures at high angles to the medial fracture suggest that movement subsequent to the lateral folding of the solidified lava skin occurred. Both compressional and tensional forces would be expected in rigid blocks "cemented" to a plastic substratum. Often, an upsurge of lava from below will flood the median fracture of a tumulus to overflow the surface or just "bead" up within it. Tumuli may also (rarely) become the loci for a hornito. Photo: Geophys. Lab., Carnegie Inst., Washington, D.C.

PLATES 130A and 130B
Two views of a pressure ridge, McCarty's Lava Flow, New Mexico. Both the longitudinal view (A) and the transverse view (B) show the medial crack which runs along the crest of the ridge. The medial cracks of pressure ridges may be filled with secondary lava. Photos: R. L. Nichols.

PLATE 131A

A pressure ridge formed on the floor of Kilauea, Island of Hawaii. Exposed at the base of the wall (background) is a lens of fine volcanic ash (visible dimension 65 by 6 meters) which was then covered by later flows, prior to formation of the Kilauea Caldera. Photo: Geophys. Lab., Carnegie Inst., Washington, D.C.

PLATE 131B

Pressure ridge with well-developed longitudinal crest fracture, Isla Fernandina, Galapagos Islands, Ecuador. Lateral compression with later internal hydrostatic pressure of subsurface lava produces the features shown. First the ridge and associated fracture formed along its crest and then the pahoehoe lava extruded over part of the tumuli surface. Banding is visible in the wall of the fracture. Photo: T. Nichols.

PLATE 132A

Rotation of lava plates, McCarty's Lava Flow, New Mexico. The photo has been turned 90 degrees counterclockwise. Grooving, tension fracturing and sharkstooth projections are well shown. Photo: R. L. Nichols.

PLATE 132B

Lava slickensides, McCarty's Lava Flow, New Mexico. The grooves on these blocks of broken pahoehoe crust were formed when the bottoms of blocks still soft and plastic were thrust across the irregular solid edges of other blocks. Photo: R. L. Nichols.

PLATE 133A

Alternating bands of glassy basaltic lava and duller, less glassy lava on the walls of a large crack in the McCarty's Lava Flow, New Mexico. The bands are about 2 to 10 centimeters wide and many meters in length but the banding is superficial, being confined to the outer 3–4 millimeters of the crack face. The bands are produced by refusion of a thin veneer on the wall as hot gases are emitted during progressive deepening of the crack. Photo: R. L. Nichols.

PLATE 133B

Sharkstooth projections on the walls of a crack in the McCarty's Lava Flow, New Mexico. The projections were formed by the scraping of the walls of the crack against one another when the lava was jtill stiffly viscous. Photo: R. L. Nichols.

PLATE 134A

Several lava blisters (Hawaii), nearly circular in outline, formed by puffing up of a thin crust of pahoe-hoe lava by gases released from the still viscous lava beneath. The blisters are hollow below the shells (1–2 centimeters thick). Most blisters in this group are only 30 to 50 centimeters wide at their bases. Notice the wrinkles in the blister at the lower right, produced by flow of the still soft crust down the slope developed during upbulging. Photo: G. A. MacDonald, U.S. Geol. Surv.

PLATE 134B

Ejecta blocks and bulbous squeeze-ups on smooth lava surface within the King's Bowl lava field in Idaho. Stony rises, part of a breached barrier dam, are in background. Caption: M. Prinz. Photo: J. Papadakis.

PLATE 135A

Air photo showing the surface of a Quaternary pantelleritic ignimbrite. Postdepositional degassing has bl stered the top surface of the unit which is now covered by blisters. Smaller examples remained intact but larger types generally collapsed to produce crater-like features. The examples of blisters shown here are approximately 100 meters in diameter. This area is on the south side of Fantale Volcano in western Ethiopia. Caption and Photo: I. Gibson, Imperial Highway Dept. of Ethiopia.

PLATE 135B

Ring structure of uncertain origin (possibly a collapsed lava blister) in the Koae fault system, Kilauea Volcano, Hawaii. The ring is about 60 meters in diameter and is composed of rubbly blocks of lava identical to that of the surrounding flow. Faults of the Koae system cut the ring structure. Caption and Photo: R. S. Fiske, U.S. Geol. Surv.

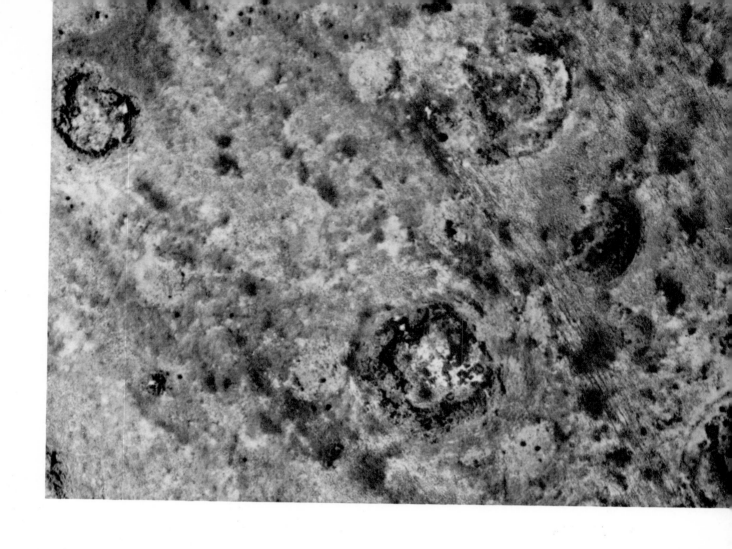

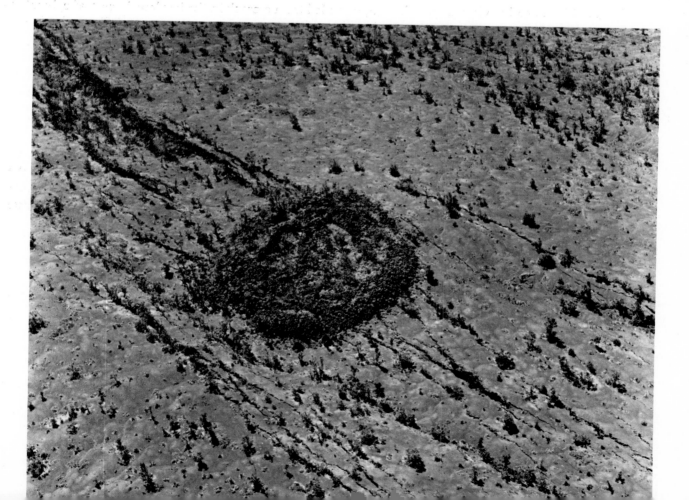

PLATE 136A

A curved lava bead of viscous basaltic lava, extruded from a crack in the hardened crust of a basaltic flow on the northwest flank of Cerro Azul Caldera, Galapagos Islands. Caption and Photo: J. Green.

PLATE 136B

A triangular wedge of fluted and grooved lava found in the medial crack of a pressure ridge on the McCarty's Lava Flow, New Mexico. The wedge formed as plastic lava was pushed up through the progressively widening medial crack in the ridge; the vertical grooves were impressed by irregularities of the wall rock at the bottom of the crack. Photo: R. L. Nichols.

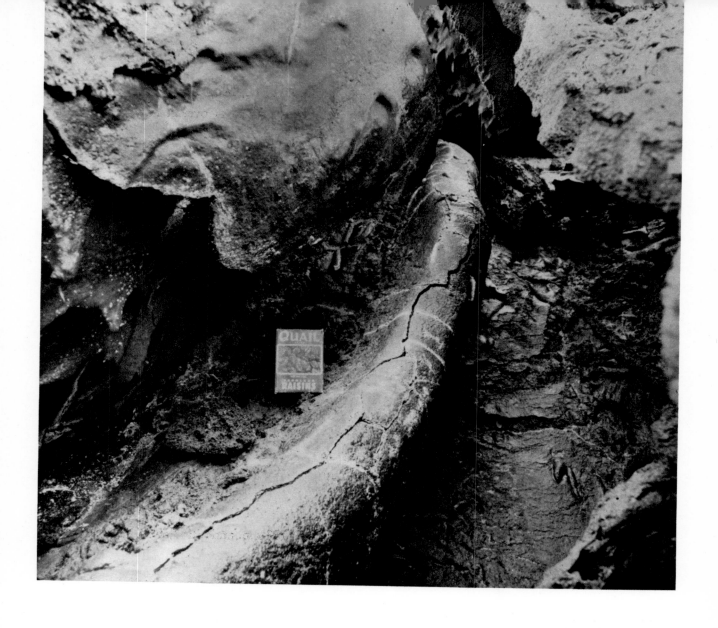

PLATE 137A

Big Obsidian Flow, Newberry Crater, Oregon. The rough jagged surface consisting of obsidian and pumice also shows numerous spines of glass extruded in the process of overall flow. The obsidian emanated from a dome to the south. Photo: R. L. Nichols.

PLATE 137B

A small lava spine (center) extruded from aa lava from the 1783 A.D. andesitic flow at Asama Volcano in Japan. Caption and Photo: S. Aramaki.

PLATE 138A

Lava stalactites, Surveyor Ice Cave, near Bend, Oregon. The ebullition of the lava river in this lava tube threw lava spatter on the cave roof which produced the "lavacicles," some of which are over 30 centimeters long. Some have blunted tips suggesting remelting of earlier spatter. Photo: Geophys. Lab., Carnegie Inst., Washington, D.C.

PLATE 138B

Lava stalactites formed in a lava tube in Hawaii by fusion through gas-heating of the upper walls and roof. Some of these slender stalactites are up to 20 to 30 centimeters in length. Shapes resemble smooth rods (right), contorted worms (center) or bunches of grapes (left). Photo: Geophys. Lab., Carnegie Inst., Washington, D.C.

PLATE 139A

Blocky lava from Cinder Cone 16 kilometers northeast of Lassen Peak, California. The extremely rough surface of the basaltic lava associated with Cinder Cone is in contrast to the smooth ash dunes on which it encroaches. The lava flow surface is about 30 meters above the surrounding terrain. The basalts are vesicular and contain quartz—a feature of particular interest to petrologists. Likewise, pumiceous inclusions in the basalt are chemically similar to the Lassen Peak dacites. Cinder Cone formed in 1851, growing to a height of 195 meters above its base; its slope angle is from 30 to 37 degrees. Large volcanic bombs are scattered around its base. Initially pyroclastic eruptions built up the cone. A lava flow from the base of the cone was followed by another series of pyroclastic eruptions. Next, a period of quiet, which was ended by further eruptions of lava from the base of the cone. Photo: G. Grant, Nat. Park Serv., U.S. Dept. of the Interior.

PLATE 139B

Block lava of the Mitsudake flow, North Yatsugatake Volcano, Japan. Caption and Photo: H. Takeshita.

PLATE 140A

An aa lava surface on the flank of Vesuvius in Italy. In the "aging" of a fresh aa field, masses of volcanic clinker are broken to fill in the otherwise more chaotic terrain. Some craggy protrusions are still present in this area. Photo: Geophys. Lab., Carnegie Inst., Washington, D.C.

PLATE 140B

Close-up of clinkery aa lava flow surface, on Isla Fernandina, Galapagos Islands, Ecuador. Although each fragment-like mass appears to be loose, the pieces are actually welded to more continuous material below the surface and are joined at contacts to each other. The width of view in the picture is about 1 meter. Caption and Photo: T. Nichols.

PLATE 141A

Edge of a pahoehoe lava lake (left) which filled a depression in rough aa lava, Isla Fernandina, Galapagos Islands. Caption and Photo: T. Nichols.

PLATE 141B

An aa lava "moraine" between two frozen rivers of pahoehoe lava on the floor of Halemaumau pit crater of Kilauea Caldera, Island of Hawaii. The river on the left is overlapped by the aa which, in turn, appears overlapped by the channel on the right. Slab and blister pahoehoe lie behind these rivers. Photo. Geophys. Lab., Carnegie Inst., Washington, D.C.

PLATE 142A

Surface of solidified lava lake in Lua Hou pit crater on the southwest rift of Mauna Loa, Hawaii. Ooze-ups have produced linear bands of lighter gray lava and more extensive overturning of the crust has renewed larger patches of the extremely fluid lava surface. Caption and Photo: R. S. Fiske, U.S. Geol. Surv.

PLATE 142B

Lava from December, 1965 eruption on the floor of Aloi Crater, Kilauea Volcano, Hawaii. During the process of crustal foundering, lava on the left side of the photo welled to the surface and is overriding slightly older crust. Drainback encrustation appears on crater walls. Caption and Photo: R. S. Fiske, U.S. Geol. Surv.

PLATE 143A

Aerial view of a lava flow over a gently sloping ground surface, Pinacate Volcanic Field, Sonora, Mexico. The hummocks and ridges, aligned normal to the direction of flow, produce a distinctive surface structure which has been called "elephant-hide" pahoehoe. Photo: T. Nichols.

PLATE 143B

Smooth, featureless surface of slab lava which was dammed, giving the surface the smooth crust of a lava lake, King's Bowl lava field, Idaho. Caption: M. Prinz. Photo: J. Papadakis.

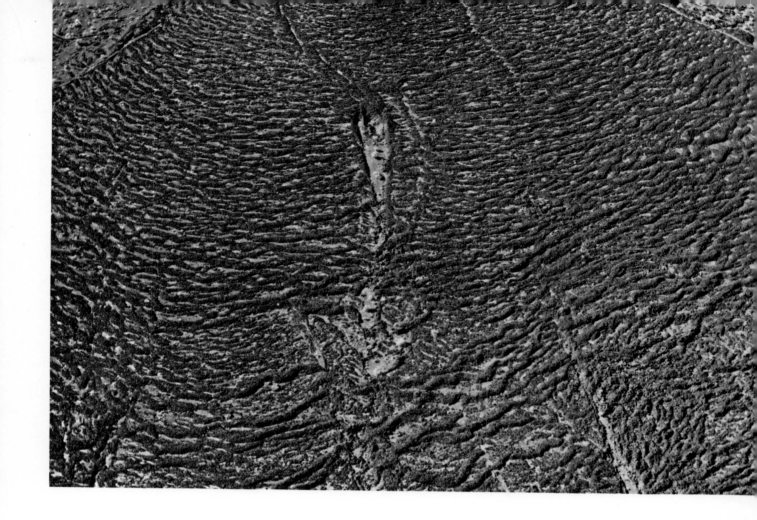

PLATE 144A

Pahoehoe lava of prehistoric age, south coast, Island of Hawaii. Pahoehoe ropes are a superficial feature grading downward to more massive basalt. Vesicles within ropes are stretched parallel to their edges. Those beneath the ropes are stretched parallel to flow direction; farther down they are circular, more numerous and larger. These relations indicate that flow was by differential laminar shear prior to and during freezing. Vesicles continued to form within its interior after flowage ceased. Caption and Photo: R. V. Fisher.

PLATE 144B

An area on the McCarty's Lava Flow, Valencia County, New Mexico in which the pahoehoe crust is broken into a jumbled but superficial mass of blocks and fragments, resembling somewhat a blocky aa field. The ruptures are shallow and this zone of disturbance probably extends a meter or so in depth. At the time of break-up, the crust over the still fluid lava was about 30 centimeters thick. The lava underneath pushed up the crust, disrupted it along cracks, and moved the broken blocks about like ice in a jam. Photo: R. L. Nichols.

PLATE 145A

Pahoehoe "toes" of olivine basalt extruded from the 1881 flow at Kilauea, Hawaii. The shape and extent of the lava lobes is dependent on the viscosity and "head" of the lava in the feeder flow. Photo: Geophys. Lab., Carnegie Inst., Washington, D.C.

PLATE 145B

Spiny or sharkskin pahoehoe, Hawaii. A pulling apart of molten to plastic lava surfaces draws the melt like taffy into the rough surface shown. The quickly chilled spines in basaltic lava are basalt glass or tachylite. Photo: G. A. MacDonald, U.S. Geol. Surv.

PLATE 146A

Festooned pahoehoe toes of Recent age, James Island, Galapagos. The basalt of low viscosity extrudes from advancing lobes of lava and is fed by the constrained reservoir of lava in the unit lobe. The pattern produced on the toe reflects the more rapid subcrustal movement of the underlying liquid lava by throwing the skin of the toe in folds convex in the direction of movement. Quick chilling of a degassing pahoehoe surface often produces the glassy pitted to hackly surface texture shown. Photo: T. Nichols.

PLATE 146B

Floor of Kilauea Caldera, Hawaii. Pahoehoe lava thrown in folds by subsurface drag effects is shown in the middle of the frozen lava stream on the left. On the right, continued passive extrusion of entrail lava has covered a small spatter cone. The caldera floor is 4 kilometers long, 3.2 kilometers wide, and 120 meters deep at its western edge. Steam vents are abundant on the floor of Kilauea as well as in the region between the principal central depression of the caldera and the outermost boundary faults at the northeastern edge of the caldera. Photo: Geophys. Lab., Carnegie Inst., Washington, D.C

PLATE 146C

Corded and convoluted pahoehoe lava, Isla Fernandina, Galapagos Islands, Ecuador. The slopes of Isla Fernandina provide a wide range in basaltic lava forms, fabrics and textures. Within a square kilometer, one can pass from a bicycle-smooth pahoehoe lava to incredibly jagged, fissured aa lava to block lava often with automobile-sized crusted autoliths to coiled and folded pahoehoe of the type shown. The color changes are as dramatic, from dark grays to chocolate browns especially in one area 4 kilometers west of Espinoza Bay. Photo: T. Nichols.

PLATE 146D

A small mass of contorted pahoehoe (Mauna Loa Flow, Hawaii). The intricate wrinkles and contortions have various aspects of ropy, corded and entrail types of pahoehoe. In the upper center, the cords seem to be elongate in the direction of flow whereas they appear to be parallel to (convex forward) the advancing front of the flow (bottom of the photo). Photo: H. G. Stephens and T. J. W. Lee III, U.S. Geol. Surv.

PLATE 147A–B

Flow surface features near Sullivan Bay, James Island, Galapagos. The photos in (A) and (B) show what appear to be the surface expressions of small spiracles (tubular openings or chimneys of varying dimensions, caused by explosive escape of gases from the lava, in which some lava may be carried upwards to the surface to pile up around the narrow orifice).

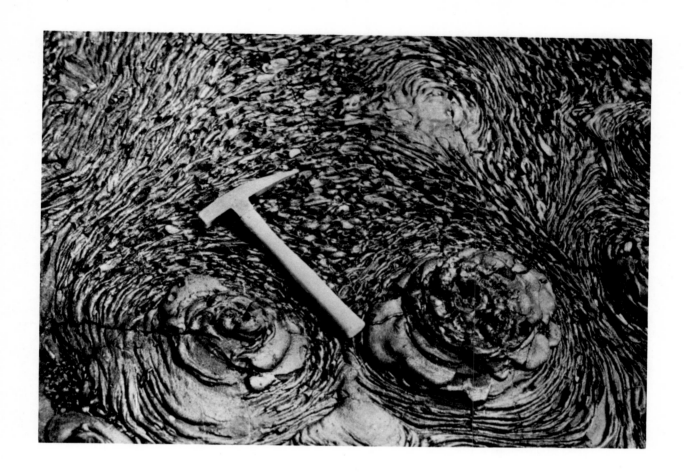

PLATE 147C–D

In (C) the direction of flow movement of the ropy lava was toward the camera, and a small lava coil has formed on the left side. (D) shows a more complex wrinkling in which surface ridges were developed parallel, as well as transverse, to the direction of flow. Caption and Photo: T. Simkin.

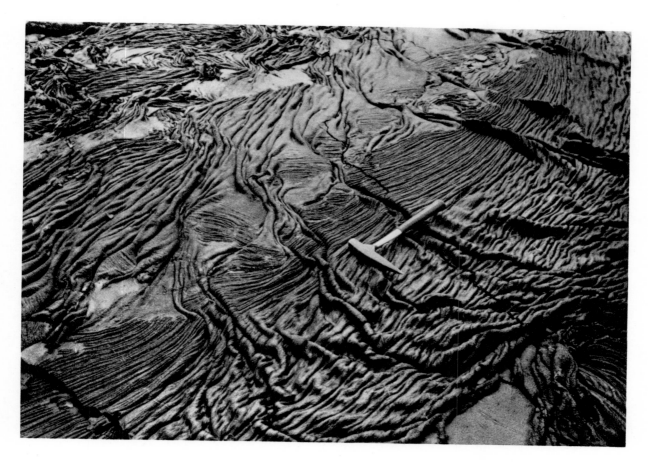

PLATE 148A

Basaltic spatter from Recent vents on north flank of Cerro Azul Caldera, Galapagos Islands. The agglutinating mechanism of spattered basalt produces much steeper slope angles of cones in the Galapagos than would be expected from viscosities measured in the lava flow channel. This paradox results from the reduction in the viscosity of the basaltic lava as it is thrown into the air. The surface of spatter is extremely vitreous and often hackly. Caption and Photo: J. Green.

PLATE 148B

Black, glassy spatter from the dwindling phases of the December 1965 eruption of Kilauea Volcano, Hawaii, lying on the surface of a lava flow erupted earlier in the same eruption. Flow lineation on the filamented lava surface is parallel to the pencil. Caption and Photo: R. S. Fiske, U.S. Geol. Surv.

PLATE 149A

Lava cascade, southwest rift zone, Kilauea, Hawaii from a photograph taken in 1920. An interesting dendritic or tree-branch structure is shown in olivine basaltic lava flowing over a small scarp. The lava flares out into a diverging flow pattern. The escarpment was caused by the sagging of one portion of the terrain relative to the other, probably due to subsurface draining of lava. The eruption (Dec. 21, 1919–Aug. 30, 1920) of which this feature is a result, was one of the largest from the southwest rift zone in this century. Photo: G. Macdonald, U.S. Geol. Surv.

PLATE 149B

Spalling of the glassy surface on a prehistoric pahoehoe flow on Mauna Loa, Island of Hawaii. Mechanical weathering, apparently frost action, begins to disintegrate the smooth flow surface into a broken mosaic of thin slabs around 5 to 10 centimeters on a side. Caption and photo: R. Decker.

PLATE 149C

A lava coil on the surface of Alae Lava Lake, Hawaii. This coplanar whorl formed at a shear zone from a strip of the thin plastic crust of the lake. The strip was progressively coiled as a result of differential flowage of the underlying molten lava. Caption and Photo: D. Peck, U.S. Geol. Surv.

PLATE 149D

Twisted, ribbon-like lava spatter lying on the surface of Alae Lava Lake, Hawaii. Caption and Photo: D. Peck, U.S. Geol. Surv.

PLATE 150A

Columnar joints in basalt at the Giant's Causeway in northern Ireland Antrim, Antrim County. The columns are here exposed in three-dimensional relief by wave erosion; in a nearby seacliff, more of the flow unit, containing columns over 10 meters in length, is visible. Note the horizontal joints spaced a third to a half meter apart in many of the columns; the top of most columns shown are surfaces controlled by these joints. Crown Copyright Geol. Surv. Photo: Reproduced by permission of the Comptroller, H. M. Stationery Office.

PLATE 150B

Close-up view of horizontal joint surfaces cut across basalt columns at Giant's Causeway, Ireland. The resulting "cross sections" reveal the polygonal character of the columns. Most columns tend to be hexagonal in outline but irregularities in the widths of each face lead to distortions of shape and elimination or addition of one or more faces in some columns. Jointing is across vertical joint planes enclosing mostly hexagons of basalt. Crown Copyright Geol. Surv. Photo: Reproduced by permission of the Comptroller, H. M. Stationery Office.

PLATE 151A

Wavy or "ball-and-socket" jointing in columnar basalts in Columbia Plateau flows, exposed south of Wenatchee, Washington. Caption and Photo: H. Coombs.

PLATE 151B

Columns in Basalt Cliffs, junction of Gardiner River and Obsidian Cliffs, Yellowstone National Park. Predominantly vertical joints are shown but horizontal surfaces are also present. The orientation of joints is controlled by the orientation of the cooling surface. Photo: G. A. Grant, Nat. Park Serv., U.S. Dept. Interior.

PLATE 152A

Devil's Post Pile National Monument, Madera Country, California. The extrusions of basaltic flows of Pleistocene age are resting on Mesozoic granitic rocks of the Sierra Nevada batholith. Many of the polygonal columns are over 20 meters high with the joints perpendicular to the cooling surface. The cause of peculiar horizontal waviness in the columnar faces (encircled) is unknown. A detailed account of the volcanics in this National Park is given by Huber and Rinehart (1965, 1966). Photo: Nat. Park Serv.

PLATE 152B

Columnar-joint rosettes associated with fossil fumarolic centers in Bandelier Tuff, Jemez Mountains, New Mexico. Caption and Photo: R. Bailey, U.S. Geol. Surv.

PLATE 153A

Basalt, First Watchung Mountain, Neward Group (Triassic) along Route 280, West Orange, New Jersey; height of cut 33.5 meters. Tomkieff structural sequence, consisting of lower colonnade, entablature, and partially exposed upper colonnade. Note radiating and bifurcating columnar joints and cross joints which are concentric about, and open away from, the apex of radiation, refraction of joints into the lower colonnade, and stratified sedimentary rocks at the base of the structural sequence (right of center). Caption and Photo: W. Manspeizer.

PLATE 153B

The Rock and Spindle, near St. Andrews, Fife, Scotland, a radiating columnar dike (left) exposed with a pinnacle of tuff that was once part of a volcanic neck in the Calciferous Sandstone Series. Crown Copyright Geol. Surv. Photo: B474. Reproduced by permission of the Comptroller, H. M. Stationery Office.

PLATE 154A

Northwest-trending (alkalic) basalt dike in the east interior wall of Haleakala Caldera. Dike is 1½ meters wide and shows prominent cooling joints oriented perpendicular to wall rock (now eroded away). Such jointing normal to the cooling surface is common in dikes, sills and thin, intercalated flows. Photo: J. Green.

PLATE 154B

Slabby basalt, Bodie Canyon area, Nevada. Slabby structure is developed only in the lowest portion of the Quaternary Aurora Crater basalt and is possibly related to the hydrous nature of the underlying felsic volcanics. Pocket microscope is 12 centimeters in length. Photo: J. Green.

PLATE 155A

Deep-sea pillow structures formed along the East Rift Zone of the Kilauea caldera system at a submarine depth of 4,450 meters. Area of photograph approximately 1.25 by 1.50 meters. Caption and Photo: J. G. Moore, U.S. Geol. Surv.

PLATE 155B

Pillow lava, Nemuro, Hokkaido, Japan. Intrusive sheets of alkali dolerite contain well-developed pillow structures with radial jointing in the central portion grading into columnar joints at the upper and lower contacts (not shown). Caption and photo: K. Yagi.

PLATE 155C

Pillow structures in lava flows at Pentire Head, St. Minver, Cornwall England. One of the earliest descriptions of the formation in pillows was by T. Anderson (1910) who observed lava from Matavanu in Savaii (Samoa) flowing into the sea. On flowing into the water ovoid masses of ropy lava were seen to swell and crack into bulb–like bodies with narrow necks and this would increase until they became as large as a sack or pillow. The connecting necks were sometimes long, but most often so short that the freshly formed lobes were heaped together. Remarkable movies of the formation of pillow lavas have been made of lava flows entering the sea at Surtsey. In the photograph shown, the concentric layers of lava in the pillow are filled with concentric layers of both filled and unfilled amygdales. Calcite makes up much of the interstitial filling between the pillows. In most cases however, the interstitial filling is palagonitized volcanic ash, or submarine muds. Crown Copyright Geol. Surv. Photo: A458–60. Reproduced by permission of the Comptroller, H. M. Stationery Office.

PLATE 155D

Basalt pillows surrounded by water-lain ash, Stapafell, Iceland. Note tabular lenses. Caption and Photo: R. Decker.

PLATE 156A

Pillow lava, Acicastello, Sicily. This sea cliff exposure of pre-Etna submarine lava clearly shows the concentric shells of basaltic glass rimming the pillows and the development of radial cooling joints in the pillow interior. Some of the pillows are almost a meter in length. A description of this area is given by Rittmann (1962, p. 70). Photo: Geophys. Lab., Carnegie Inst., Washington, D.C.

PLATE 156B

Pillow-shaped structures in Diamond Head tuff, Oahu, Hawaii. Road cut is on the seaward side of the Diamond Head tuff crater; view here is to the south. The tuff is now palagonitic. Caption and Photo: J. Green.

PLATE 157A

Ovoidal structures formed in basaltic (mugearite) lava near North Berwick, East Lothian, Scotland. These structures presumably form by uniform cooling from the molten condition during lava emplacement, so that stresses are relieved by spheroidal parting rather than radial cracking around cooling centers. Note the onion-skin structure (top center); the honeycomb-like septa in these structures may be fillings of secondary fractures. Crown Copyright Geol. Surv. Photo: C493–6. Reproduced by permission of the Comptroller, H. M. Stationery Office.

PLATE 157B

Roadcut through pahoehoe lava of late prehistoric age, south flank of Kilauea Volcano, Hawaii. The interior of many flow lobes drained downslope, producing numerous small lava tubes. The number of flow units in this exposure is uncertain. Caption and Photo: R.S. Fiske, U.S. Geol. Surv.

PLATE 158A

Volcanic bombs and other ejecta on the floor of Haleakala Caldera, Maui, Hawaii. Olivine-rich basaltic bomb in foreground center is 1.5 meters long and exhibits sag structures and cooling fractures on its chilled crust. This bomb was thrown an estimated 300 meters from its source cone on the caldera floor (background). The wide size range of bombs observed in this area is unusual. Photo: J. Green.

PLATE 158B

Ejected block of augite-hypersthene andesite thrown from Asama Volcano in December, 1958. The block is 4 meters in maximum direction and weighs about 20 tons. The high degree of explosivity of Asama Volcano differs from that of more basaltic type volcanoes because of the higher viscosity of the andesite magma. As a result, large blocks such as this one can be ejected at high speeds from the volcano vent. The volcano physics of this eruption is well documented by T. Minakami and his associates who showed that a swarm of shallow earthquakes preceeded the onset of the most violent phase of the 1958 eruption. Photo: S. Aramaki.

PLATE 159A

Secondary volcanic bomb impact craterlets, in basaltic ash, near Kapoho, Hawaii. Largest craterlets are 3 meters in diameter. Volcanic ejecta was from fire fountaining of February, 1960. Many of the craterlets show the blocks and bombs within them that produced the depressions which often have raised rims. The persistence of the impact craters over at least a two-year period is due to the high porosity of the cinder field and a slight surface crustiness produced by weathering and/or sublimate deposits which make the surface resistent to erosion. A description of this area is given by Hartmann (1967). Photo: L. Nichols, courtesy of Dr. G.P. Kuiper.

PLATE 159B

Detail of Kapoho volcanic bomb impact field showing close-up of 2 meter-wide impact crater. The craterlet has a poorly defined rim as does a second craterlet in upper right. Similar impact craterlets produced by volcanic ejecta are common in all volcanic provinces of the world where ejecta lands in relatively scft pyroclastic blankets. Excellent examples may be found at the volcano in the Batur Caldera in Bali, and at nearby Keanakakoi where blocks from the 1790 phreatic explosion of Halemaumau have landed in basaltic cinders. Photo: G. Kuiper.

PLATE 160A

Impact craters about 3 kilometers west of Arenal Volcano, Costa Rica. Volcanic bombs and fragments ejected during the 1968 eruption from the flank of this volcano produced these secondary craters, most of which were filled with water at the time this photo was taken. The surface of impact was largely ash-covered from earlier stages of the eruption. Caption: W. Melson. Photo: T. Simkin.

PLATE 160B

Sag effects in layered sideromelane tuff near MacDoel, California. Distortion of tuff bedding was caused by impact and settling of volcanic blocks thrown from a maar crater on the rim of which the outcrop occurs. The bedding is deformed mostly by impact and settling below the block and by differential compaction above it. Photo: H. Wilshire, U.S. Geol. Surv.

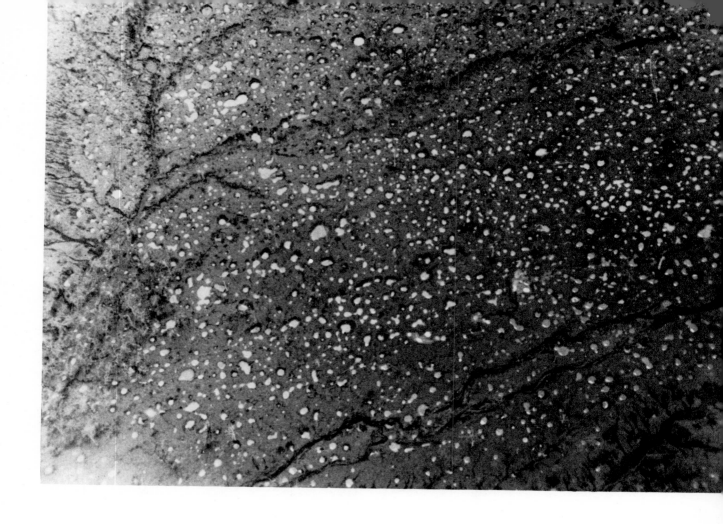

PLATE 161A

Augite-hypersthene andesite bread crust block (bomb) near summit of Asama Volcano, Japan. Caption and Photo: J.G. Moore, U.S. Geol. Surv.

PLATE 161B

Breadcrust bomb of pumice, south shore of East Lake, Newberry Caldera, Oregon. Shrinkage cracks about 0.1 to 1 centimeter wide occur on the skin of this glazed pumice bomb. The cracks typically taper inward to give a surface similar in texture to a loaf of bread. Photo: J. Green.

PLATE 161C

Aerodynamically shaped, volcanic bomb; San Francisco Mountains, Arizona. Stretch marks near the central part of the bomb to the right of the geological hammer define flow lines produced during flight. The duration of flight for most volcanic bombs is only on the order of a minute or less. Some bombs are still molten in their interiors on impact and deform or sag subsequent to landing. A comprehensive account of the morphology of volcanic bombs is given by Reck (1875). Photo: T. Nichols.

PLATE 161D

Cored volcanic bomb, Silali Volcano, Kenya. The core consists of a somewhat vesiculated gabbroic nodule; the rind is an alkalic tachylitic basalt. The rind is broken here to show the interior core. The melting point of the gabbro was either higher than that of the enveloping lava or the introduction of the gabbro to the lava in the throat of the vent was sudden. Photo: G.J.H. McCall.

PLATE 162A

Pahoehoe basalt, P. Pahala flow near Pakanaka Hill, Hawaii. Vesicular texture is well shown in this hand specimen. The marked sphericity of the vesicles and the uniformity of their spacing, however, does not usually persist much beyond the scale of a meter due to variations in cooling rate and local gas content of a flow unit. Photo: H.T. Stearns, U.S. Geol. Surv.

PLATE 162B

Basaltic scoria, Craters of the Moon National Monument, Idaho. Texture is produced by the exsolution of gases on release of pressure during a volcanic eruption. The expanding gases produce pits, the size of which is a function of the rate of pressure release, temperature of lava, viscosity of lava, initial volatile content, rate of cooling, and presence of bubble nucleation sites. Volcanic bombs are more scoriaceous on their interiors than exteriors. On the other hand, volcanic rocks carried in a nuée ardente do not show a serial change in vesicularity from rind to core. Photo: G. A. Grant, Nat. Park Serv., U.S. Dept. Interior.

PLATE 162C

Pipe vesicles in Burney Peak flow near Burney, California. These cylindrical tubes usually less than a centimeter in diameter form at the base of a lava flow and rise nearly perpendicularly from the lower margin. In some cases two pipes merge. The forks of the split pipe point toward the base of the flow. This technique can be used with caution in determining whether or not a lava flow is overturned or right side up. Pipe vesicles are related to cooling and gas expansion on the margins of lava flows with moderate viscosity and low surface tension. Photo: H. Wilshire.

PLATE 162D

A volcanic bomb in a cinder quarry west of Paulina Peak, Newberry Caldera, Oregon. The highly vesicular interior is coated with denser, less vesicular lava of the same composition. Within the highly vesicular interior, there is an increase in vesicle diameter toward the interior. Photo: J. Green.

PLATE 163A

Nodular rhyolitic lava of the Bala Volcanic series, along the Caernarvon Road, north of Pwllheli Station, Caernarvonshire, England. Crown Copyright Geol. Surv. Photo: A6505. Reproduced by permission of the Comptroller, R. M. Stationery Office.

PLATE 163B

Porphyritic texture in andesitic lava. Equant to anhedral phenocrysts of feldspar are visible. Borrowdale Volcanic Series, exposed southeast of Eycott Hill, Cumberland, England. Crown Copyright Geol. Surv. Photo: A6614. Reproduced by permission of the Comptroller, H. M. Stationery Office.

PLATE 163C

Weathered surface of a silicic spherulitic dike cutting a gabbro of the Tertiary Igneous Series at Druim an Eidhne, Cuillin Hills, Isles of Skye, Inverness-shire, Scotland. The light-colored ovoid bodies are spherulites, composed of fibrous, radiating feldspar crystals, which grew and partially filled open cavities in the dike material. Crown Copyright Geol. Surv. Photo: B163–4. Reproduced by permission of the Comptroller, H. M. Stationery Office.

PLATE 163D

Flow banded, spherulitic rhyolite at Aratiatia, New Zealand. Caption and Photo: B. Thompson.

PLATE 164A

Lithophysae, exposed in Obsidian Cliff, Yellowstone National Park, Wyoming. The largest ones are 15 to 20 centimeters in diameter. Caption and Photo: R. Decker.

PLATE 164B

Accretionary lapilli formed during the 1965 eruption of Taal Volcano (see Plate 103), Philippines. These are commonly produced in ash-charged clouds by accretion of ash around a moist core. Caption and Photo: J. G. Moore. U.S.Geol. Surv.

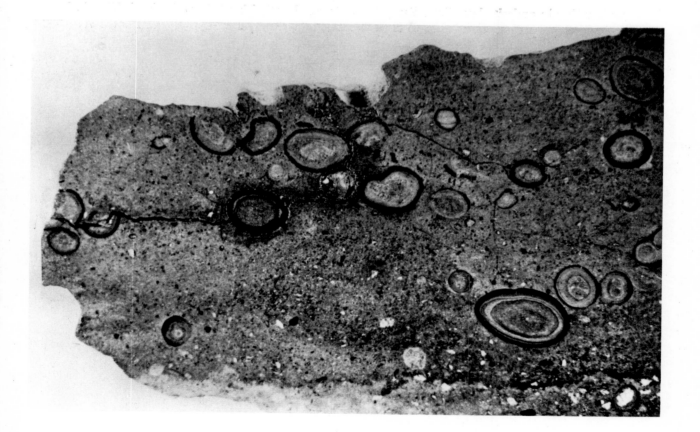

PLATE 165A

Flow banding in andesite lava flow, Ruapehu Volcano, North Island, New Zealand. The dark layers contain a higher proportion of glass to crystals than the light layers. Caption and Photo: W. Parsons.

PLATE 165B

Welded andesitic scoria flow, with xenoliths of older flow rock (jacket for scale in lower right); Ruapehu Volcano, New Zealand. Caption and Photo: W. Parsons.

PLATE 165C

Fine, closely-spaced flow-banding in fine-grained, nonporphyritic andesites of the Borrowdale Volcanic Series, outcrop south of Eycott Hill, Cumberland. England. Crown Copyright Geol. Surv. Photo: A6759. Reproduced by permission of the Comptroller, H. M. Stationery Office.

PLATE 165D

Silicic dike cutting across gabbro at Druim an Eidhne, Cuillin Hills, Skye, Scotland. This is a common association in the eroded calderas of the Tertiary province of Scotland. The higher viscosity of the dike during emplacement is well-illustrated by the intricate flow banding which is highlighted by differential erosion. Crown Copyright Geol. Surv. Photo: B162. Reproduced by permission of the Comptroller, H. M. Stationery Office.

PLATE 166A

Banded rhyolitic glass, developed during flowage of a viscous lava, exposed in a block within the central plug of the Panum Crater, Mono Craters, California. Height of exposure is about 1.5 meters. Caption and Photo: J. S. Shelton. From *Geology Illustrated* by J.S. Shelton, W.H. Freeman and Co. Copyright, 1966.

PLATE 166B

Hand specimen of banded obsidian from Mono Craters, California. Typical conchoidal fracture is clearly seen in this 12-centimeter long sample. Photo: J. Green.

PLATE 167A

Cliffs in ash flow deposits of the Bandelier Tuff as seen from New Mexico State Route 4, Los Alamos Canyon, Pajarito Plateau, Jemez Mountains, New Mexico. Flows forming darker, conspicuously jointed cliffs are more densely welded than those forming lighter slopes. Caption and Photo: R. Smith, U.S. Geol. Surv.

PLATE 167B

Ash flows of the Bandelier Tuff overlying bedded ash fall tuffs and reworked tuffs, Pueblo Canyon, Pajarito Plateau, Jemez Mountains, New Mexico. Ash flows from top of bedded tuffs up to first ledge are vitric and nonwelded; flows from ledge to top of cliffs are crystallized from vapor phase activity and show alternations in degree of welding. Caption and Photo: R. Bailey. U.S. Geol. Surv.

PLATE 168A

Bedded welded tuff in upper part of ash-flow sheet of Topopah Spring Member, Nye County, Nevada. This close view of a densely welded zone shows stratification owing to variations in size and abundance of pumice lapilli and phenocrysts. Partings in lower right half of photo are in densely welded crystallized tuff and follow concentrations of eutaxitic pumice lapilli which have preferentially weathered out. Darker layer behind hammer is a vitrophyre in which phenocrysts and pumice contents are similar to those of underlying crystallized tuff. Thin, light-colored layers above hammer head are beds in which phenocrysts and pumice contents are lower than average. Caption and Photo: P.W. Lipman, U.S. Geol. Surv.

PLATE 168B

A specimen (collected by C. Fries, Jr.) from a 10 to 20-meter thick layer of welded tuff, at kilometer 153 of Taxco Highway, State of Guerrero, Mexico. The lighter colored groundmass shows a fine-grained pyroclastic structure. The darker areas represent completely collapsed pumice fragments. The streaky appearance of this rock is termed eutaxitic structure. It is emphasized in this specimen by weathering along a joint surface; the opposing side of the specimen has an obsidian-like appearance. Caption and Photo: R. L. Smith, U.S. Geol. Surv.

PLATE 169A

Xenolithic layer in lava at Glenbyre, Isle of Mull, Argyll, Scotland. Copyright Geol. Surv. Photo: C2722. Reproduced by permission of the Comptroller, H. M. Stationery Office.

PLATE 169B

Outcrop in Black Glass Canyon showing typical appearance of the xenolithic unit in the Topopah Spring Member, Nye County, Nevada. The xenoliths are subequant and slightly rounded, fairly uniform in size, and heterogeneous in rock type. Among the more conspicuous inclusions are quartz monzonite, crystal-rich quartz latitic welded tuff, crystal-poor rhyolitic welded tuff, vesicular flow-banded rhyolitic lava, and dense crystallized flow-banded rhyolitic lava. The many other inclusions are difficult to discern from the matrix; total inclusion content is greater than 50 percent. Caption and Photo: P. W. Lipman; U.S. Geol. Surv.

PLATE 169C

Xenoliths from the 1801 flow Hualalai Volcano, Hawaii. Unusual quantities of olivine- and pyroxene-rich inclusions were laid down locally in gravel-like deposits. Most of the clasts are coated with a thin skin of glassy basalt. Caption and Photo: R. S. Fiske. U.S. Geol. Surv.

PLATE 169D

Buchite xenoliths near upper contact of Hunish Sill, northwest Trotternish, Isle of Skye, Scotland. Formerly flat–lying pelitic sedimentary rocks have been plastically deformed and metamorphosed to at least the pyroxene hornfels facies and probably to the sanidinite facies. Caption and Photo: T. Simkin.

PLATE 170A

Accretionary lava ball formed on an aa flow at Lava Butte, Oregon (see Plate 54B). Such balls develop when pyroclastic fragments are coated with viscous lava. The ball shown here is partially buried and the shell has been partly removed so that the core is visible. Lava has piled up on the upflow side. Caption and Photo: R. L. Nichols.

PLATE 170B

Detailed view of the pyroclastic core of the lava ball shown in (A) above. Caption and Photo: R. L. Nichols.

PLATE 170C

Ball or spheroidal lava, near Kilauea, Hawaii. Spheroidal lava bombs may range from a few centimeters to as much as a meter in diameter. The inner portion consists of vesicular lava, but in contrast to a normal volcanic bomb there is a discontinuity of texture between the skin and center of the bomb where finer-grained denser lava has enveloped a previously formed mass in the throat of the volcanic vent. The origin is comparable to that of a cored bomb, but in the latter the bomb core is not lithologically similar to the rind. Photo: H. T. Stearns, U.S. Geol. Surv.

PLATE 171A

Agglomerate, Puu Hou littoral cone, south coast, Island of Hawaii (see Plate 51B). Layers are agglutinated bombs near source of explosions (as in photo) where lava entered the sea. Many bombs shown have oxidized red centers and thick rinds of glass; projection of dip shows they were derived from below sea level. Caption and Photo: R. V. Fisher.

PLATE 171B

Volcanic agglomerate in the Absaroka Mountains, east of Yellowstone National Park, Wyoming. Caption and Photo: R. Decker.

PLATE 172A

A volcanic breccia in the Lower Old Red Sandstone Volcanic Series, east-northeast of Clachaig, Glen Coe, Argyll, Scotland. Occasionlly in volcanic breccias, some of the fragments are more rounded than the definition implies. This is because there is some degree of hydraulic churning which rounds many of the angular fragments. The size of fragments in the photograph shown ranges from millimeters to centimeters. Crown Copyright Geol. Surv. Photo: Reproduced by permission of the Comptroller, H. M. Stationery Office.

PLATE 172B

Wairakau Breccias, Whangaroa Harbour, New Zealand. A volcanic breccia of brecciated flow fragments in lower part of photograph shows a gradational contact to tuff breccia of subaqueous origin in the upper two-thirds of the picture. Groundmass of the tuff breccia is a mixture of fine angular crystal and lithic debris and represents over 75 percent of the total deposit. Caption and Photo: W. Parsons.

PLATE 172C

Breccia at top of a late Pleistocene rhyolite flow of the Madison Plateau, in southeastern Idaho just west of Yellowstone National Park. Blocks of black obsidian lie in an unconsolidated matrix of shards, strands, and granules of clear glass. The exposure is about 7 meters high. Caption and Photo: W. Hamilton, U.S. Geol. Surv.

PLATE 172D

Monolithologic volcanic breccia in the Absaroka Mountains, Wyoming. Caption and Photo: R. Decker.

PLATE 173A

A mound (erosional outlier) of basaltic pillow breccia and palagonite, near Ingólfsfjall, Iceland. Caption and Photo: R. Decker.

PLATE 173B

Cross section of typical cinder cone material, Pukeonake cinder cone, Tongariro National Park, North Island, New Zealand. Note initial dip (top edge of picture is the horizontal), reverse graded bedding, subangular character of cinders and bombs, and almost complete lack of groundmass in coarse cinder layers. Caption and Photo: W. Parsons.

PLATE 173C

Autobreccia of cellular glass in phonolitic trachyte froth-flow, Menengai, Kenya. Caption and Photo: G. J. H. McCall.

PLATE 173D

Clastic dike consisting of a mosaic of angular andesitic fragments in a matrix of marine limestone. The andesite was emplaced as a near-surface flow which intruded beneath limy sediments on the floor of the middle Miocene sea. Exposure in Palos Verdes Hills, southern California. Caption and Photo: G. Eaton.

PLATE 174A

Avalanche breccia of slightly vesicular, perlitic rhyolite blocks and ash on western slopes of Tuahu rhyolite dome near Atiamuri, New Zealand. Blocks are up to 0.7-meter in diameter. Caption and Photo: B. Thompson.

PLATE 174B

Late Quaternary rhyolitic pumice breccia, deposited from a pumice avalanche. Pumice fragments up to 3 centimeters in length. Deposit occurs near Wairakei, New Zealand. Caption and Photo: B. Thompson.

PLATE 174C

Laharic volcanic breccia of Recent age is shown with angular to subangular blocks of andesite in a groundmass of ash, silt, and clayey material, National Park highway cut near the Chateau, New Zealand. Caption and Photo: W. Parsons.

PLATE 174D

Ice-contact palagonite breccia, Coulman Island, Antarctica. Chunks, nodules, and spheroids of vesicular basalt lie in a matrix of granules of hydrated basalt glass. The complex formed in Pliocene time when grounded ice filled the Ross Sea, which is now largely ice-free. Caption and Photo: W. Hamilton, U.S. Geol. Surv.

PLATE 175A

Pumice flow deposits, the Taupo Pumice, with no bedding or sorting. Note charcoal fragments and reworked material at top; road cut is in National Park highway, near Tongariro National Park, North Island, New Zealand. Caption and Photo: W. Parsons.

PLATE 175B

Finely stratified, moderately consolidated pumice breccia, probably deposited in water; near Ohakuri, New Zealand. Caption and Photo: B. Thompson.

PLATE 175C

Embankment exposing an airfall pumice deposited during the eruption of Haruma (Japan) volcano about 1,300 years ago. Caption and Photo: S. Aramaki.

PLATE 175D

Airfall ash and pumice deposits, showing good lamination and sorting in fine and coarse ash and pumice lapill beds, from post–caldera explosive volcanism, in Valles Caldera, New Mexico. Caption and Photo: W. Parsons.

PLATE 177A

Mantle-bedded pumice lapilli beds, El Cajete Member, Valles Rhyolite, Jemez Mountains, New Mexico. The anticline-like structure results from air-fall mantling of the deposits over a small hillock in the underlying topography. Caption and Photo: R. Bailey, U.S. Geol. Surv.

PLATE 177B

Pleistocene and recent airfall tuffs of basaltic composition on the flanks of O Shima Volcano, Japan. The lower sequence of beds rests upon and mimics an irregular erosion surface lying below the level of the road. Rapid erosion carved a new, irregular surface into this sequence and later eruption showered debris onto the surface. None of these strata have been folded; all dips are primary. Caption. R.S. Fiske. Photo: J. Green.

PLATE 179A

Erosion furrows caused by heavy rainfalls acting on unconsolidated ash at the base of the cone of Parícutin Volcano, Mexico. Caption and Photo: T. Nichols.

PLATE 179B

Batok Volcano within the Tengger Caldera, Java. In the "Sandsea" of the Tengger there are seven eruptive centers located on two intersecting zones; one east-west and the other NNE by SSW. Batok, about 150 meters high, shows conspicuous volcano flank gulleys eroded by the heavy rainfall of the area. The furrows were initiated by the rolling down of volcanic bombs and blocks produced by the eruptions. The ejecta produced is mainly hypersthene basalt. Photo: R. Decker.

PLATE 181A

Trees denuded by a pumice-scoria fall during formation of the cinder cone in the center background. This cone formed in 1959 along the edge of Kilauea Iki pit crater (see Plate 124A). Caption and Photo: R. S. Fiske, U.S. Geol. Surv.

PLATE 181B

Impact craters about 4 kilometers from explosion crater on Arenal Volcano, Costa Rica. Base surge following blast associated with explosion (during 1968 eruption) defoliated and debranched trees in background. Caption and Photo: W. Melson.

PLATE 183A

Another detail of preservation of a charred wood pattern preserved in 'ava at Craters of the Moon National Monument, Idaho. Evidently a negligible amount of differential movement occurred after the charring of the wood. Photo: W. Keller, Nat. Park Serv., U.S. Dept. Interior.

PLATE 183B

Cross section through a tree mold in aa lava field, Lava Cast Forest, Oregon. Caption and Photo: R. L. Nichols.

PLATE 184A

Horizontal tree mold showing impression of the bark, at Craters of the Moon National Monument. Idaho. Photo: R. Menning, Nat. Park Serv., U.S. Dept. of the Interior.

PLATE 184B

Tree rings preserved in lava, Lava Cast Forest, Oregon. Over 100 rings are preserved in the specimen shown. Note the change in the width of the tree rings. Photo: R. L. Nichols.

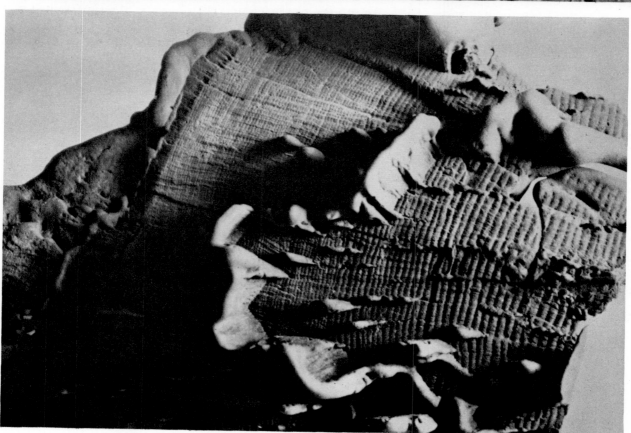

PLATE 185A–B

Eruption sequence of Strokkur Geyser in the classical thermal spring area of Haukadalur, southern Iceland, August 6, 1968. Emergence of a bubble (A) one to two meters in diameter, as a large mass of superheated water rises from a central tube, marks the onset of eruption from a circular geyserite basin filled with near-boiling water. Explosive rupture of the bubble (B) occurs as the superheated water changes to steam.

PLATE 185C–D

This change is accompanied by outward movement of a well-defined circular wave (C) in the geyser pool and is followed by eruption of a geyser column (D) 20 meters or more in height. A strongly perceptible seismic pulse occurs almost simultaneously with rupture of the bubble. The water level of the geyser pool is lowered several centimeters after each eruption as a result of outflow down a natural drainage channel (foreground, C) as the wave reaches the perimeter of the pool and additional back flow of eruption water down the central tube. The Strokkur eruption sequence is followed by a series of seismic pulses, beginning several seconds after the eruption; sloshing water filling the emptied underground cavities probably generates the pulses. The eruptions of Aug. 6, 1968 occurred at 10 to 15-minute intervals. Caption: J. D. Friedman, U.S. Geol. Surv. Photo Sequence: Kerstin Friedman.

PLATE 186A

Pohutu geyser, Whakarewarewa, Rotarua Thermal District, North Island, New Zealand. This geyser which plays from one to four times a week produces steam and water jets about 20 meters high. A "sympathetic" geysering action also occurs at these times in the nearby Prince of Wales Feathers geyser. There are about eight geysers in this thermal area which have built up extensive sinter terraces. Photo: T. Ulyatt, New Zealand Geol. Surv.

PLATE 186B

Old Faithful Geyser, Upper Basin, Yellowstone National Park, Wyoming. This most famous of America's geysers erupts approximately every hour to heights of 40 meters or more, discharging on average about 10,000 to 12,000 gallons of steam and water. Eruption heights vary somewhat with the amounts of rainfall in the season. Photo: J. S. Rinehart.

PLATE 187A

Liberty Cap hot spring deposit of travertine, near Mammoth Hot Springs, Yellowstone National Park. Abundant hot springs occur in this northern part of the park. Some have built up travertine cones by the precipitation of calcium carbonate salts from water that traverses the underlying limestone. The silica content of the water is quite low. Water temperatures are generally lower than elsewhere in the park and lowest on the upper terraces. The maximum temperature recorded is 75°C. These low temperatures account for the absence of geysers. Many of the springs are ephemeral leaving extinct cones of the type shown to indicate their former activity. Photo: Geophys. Lab., Carnegie Inst., Washington, D. C.

PLATE 187B

The Grotto Geyser, Upper Geyser Basin, Yellowstone National Park. Photo: J. Hansen, Nat. Park Serv., Dept. of the Interior.

PLATE 188A

The Lone Star Geyser in the Yellowstone Basin, Yellowstone National Park, Wyoming. The geyser, built of siliceous sinter, erupts every one to two hours for a 10-to 20-minute period, with its plume reaching heights up to 15 meters. The hot waters emitted are alkaline with considerable silica as well as bicarbonates, carbonates, chlorides and fluorides of the alkali metals presumably derived by the leaching of the rhyolitic rocks underlying Yellowstone. These rocks consist of thick accumulations of flows, pumice breccias and welded tuffs which bury a highly irregular topography developed on an even thicker accumulation of early and middle Tertiary intermediate and basic lavas, breccias, tuffs, agglomerates and mudflows. For a discussion of the chemistry of the Yellowstone waters as they relate to the underlying rocks, see White (1957). Photo: Geophys. Lab., Carnegie Inst., Washington, D.C.

PLATE 188B

Lone Star Geyser during a period of quiescence. The multiple holes through which the steam and water are jetted are well shown. Photo: Geophys. Lab. Carnegie Inst. Washington, D. C.

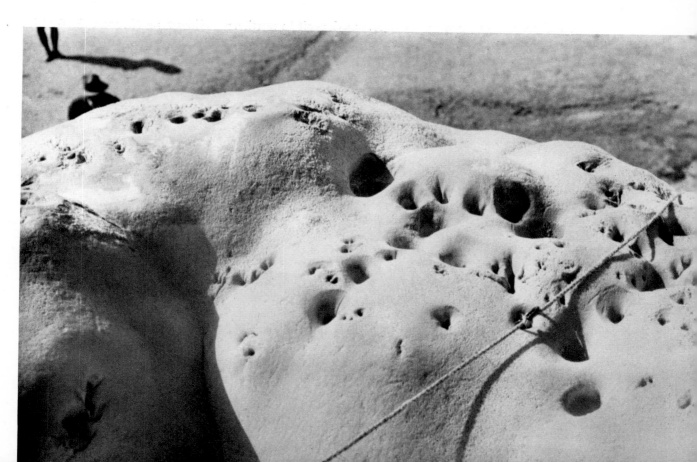

PLATE 189A

Mammoth Hot Springs, calcareous terrace deposits at the north end of Yellowstone National Park, Wyoming. Terraces originate by precipitation of calcium carbonate (travertine). Photo: Nat. Park Serv., U.S. Dept. of the Interior.

PLATE 189B

Jupiter Terrace Hot Springs, a calcareous deposit in northern Yellowstone National Park, Wyoming. Photo: C. Hansen, Nat. Park Serv., U.S. Dept. of the Interior.

PLATE 190A

A small spring with ornate deposits of sinter (SiO$_2$) 50 meters east of Union Geyser, Yellowstone National Park, Wyoming. Caption and Photo: D. E. White, U.S. Geol. Surv.

PLATE 190B

Siliceous sinter in Porkchop Spring, Norris Basin, Yellowstone National Park, deposited from alkaline waters of high silica content. Note differences in sinter above and below water level, and indications of current directions. Caption and Photo: D. E. White, U.S. Geol. Surv.

PLATE 191A

"Boiling" mud pot, Norris Basin, Yellowstone National Park, Wyoming. Often a great deal of relatively cool entrained gases provide much of the turbulence to "boiling" mud springs. Norris Basin, about 3 square kilometers in area, also has 34 active geysers, many hot springs, and some superheated steam jets. Most of the waters are sulfate-chloride in nature, neither dominantly acid nor alkaline. Spring deposits are principally siliceous sinter. Photo: N. Short.

PLATE 191B

Small "spatter cone" vent of mud volcano about 20 centimeters high with extrusion of mud flow, at Pocket Basin, Lower Basin, Yellowstone National Park, Wyoming. Caption and Photo: D. E. White, U.S. Geol. Surv.

PLATE 192A

Bumpass Hell area, 3 kilometers south of Lassen Peak, California. This crater-shaped basin is a zone of hot springs, fumaroles, and mud pots, active prior to the 1915 eruptions of Lassen. Hydrothermal alteration of rock and soil leaves the ground of Bumpass Hell weak, hollow to the tread, and easily broken through. Acid leaching produces siliceous residue and kaolinite. Springs are all acid-sulfate, with H_2SO_4 derived from hydration of H_2S. Photo: A. L. Day.

PLATE 192B

Solfatara groups, thermal springs, and steaming ground of the rhyolitic Hrafntinnusker area (foreground), part of the larger Torfajökull volcano–tectonic system, south-central Iceland, August 15, 1969. Nearly all hydrothermal features of the Torfajökull region, the largest geothermal field in Iceland, lie within the apparent rim of a subsided caldera in the Torfajökull rhyolite massif. This inferred ring fracture encloses the major postglacial and historic silicic volcanic center in Iceland. Caption and Photo: J. D. Friedman, U.S. Geol. Surv.

PLATE 193A

Deposition of sulphur along a steaming fissure from which lava erupted in 1950, southwest rift of Mauna Loa, Hawaii. Caption and Photo: R. S. Fiske, U. S. Geol. Surv.

PLATE 193B

Sulphur encrusting a broken and tilted tree mold and surrounding lava surface along the East Rift Zone of Kilauea Volcano, Hawaii; initial eruptions took place in December 1965 but the photo was taken in 1967. Caption and Photo: R. S. Fiske, U.S. Geol. Surv.

PLATE 194A

White and yellow encrustations and fillings of salts (sulphur?) depositing on the north side of Vulcano, Lipari Islands, where no rain had fallen for 8 months prior to date of photo (Aug. 20, 1914). Photo: Geophys. Lab., Carnegie Inst., Washington, D.C.

PLATE 194B
Rosette crystals of sulphur deposited on rock exposed in the crater wall of Vulcano, Lipari Islands, north of Sicily. Photo: Geophys. Lab., Carnegie Inst., Washington, D.C.

PLATE 195A

Lunar Orbiter V, Site 38, Frame 159, Medium Resolution (∼30 meters) photograph of a region in the center of Mare Imbrium near the crater Carlini at 23° W, 33° N (lunar coordinates). The major feature of the photograph is an apparent flow unit whose lobate front can be distinguished. The two largest craters are Carlini (upper left) and Carlini B (lower right), the latter being 10 kilometers in diameter. The flow front has a scarp that may be approximately 100 meters high in some places. It can be seen in earth-based photographs at high sun angles as a discontinuity in color contrast. Other color differences have also been shown on Orbiter pictures to be discontinuities in elevations, indicating other flow units. Such elevation differences had been searched for by earth-based techniques but had not been detected until shortly before the use of Orbiters. The lobate or scalloped edge and presence of a terminal scarp are characteristic of many terrestrial lava flows. The size of this flow is comparable to large lava plateaus on earth. If the composition of the lunar flows is basaltic, they may be similar to our extensive flood basalt plains such as the Columbia River Plateau in the northwestern United States. The Surveyor and Apollo 11 and 12 analyses on other maria favor basaltic lava. A few small furrows occur on the flow surfaces. Caption: W. Cameron. Photo: NASA Langley Research Center.

PLATE 195B

Luanr Orbiter IV Site 33D, Frame 189 (2nd of 3), High Resolution (∼2 meters) photograph centered approximately at 80° W, 40° N. Most of these craters are unnamed but the large one near the right edge is Lavoisier (65 kilometers in diameter). This region is rather unusual as it contains so many craters that are highly cracked or rilled. Certain features in this photograph suggest volcanism. A small crater in the center of Lavoisier and two larger ones at the extreme top center of the photograph have double concentric walls that are not likely to be chance successive bull's eye impacts. Instead, they may be instances of viscous lavas that extruded along ring fractures, similar to ring–dike features on earth. In Lavoisier there are many concentric rills, sometimes alternating with concentric ridges and occasionally converting from one to the other, as at the western (left) edge of Lavoisier. Similar occurrences are found in the other craters with numerous rills. Caption: W. Cameron. Photo: NASA Langley Research Center.

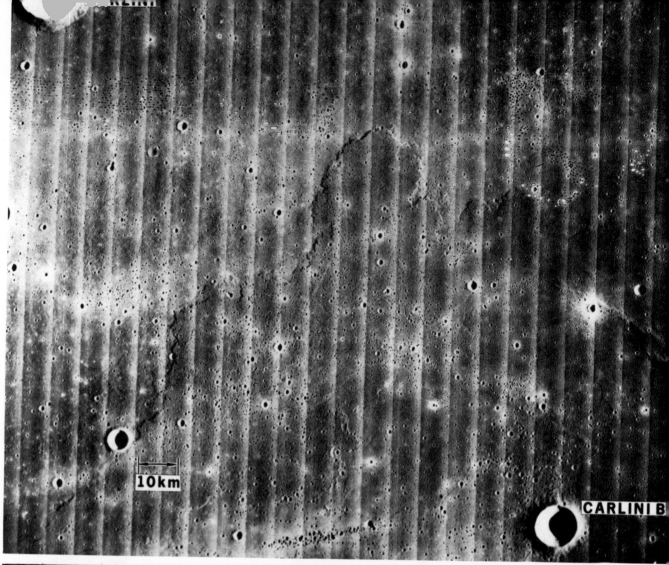

10km

CARLINI B

NORTH

65 km

LAVOSIER

PLATE 196A

Lunar Orbiter II, Frame 213, Medium Resolution (~30 meters), oblique view of part of the field of Marius domes. The large (45 kilometers) crater at the top is Marius (51° W, 12° N). The radiating ridges and elongated craters at the left edge are from the crater Reiner. This is the largest dome field on the Moon and is composed of more than 100, varying in size and structure. Note the frequent association with low, linear ridges. In the center is one of the largest (~18 kilometers) ridges, somewhat irregular in shape and part of a complex of ridges. It is highly cratered; whereas the smaller dome to the left of its far end is a symmetrically round dome with only one summit crater. Others have no craters on them, and still others have small peaks of steeper slope than those of the main part having 1–5° slopes. The heights of the Marius domes range from tens to a thousand meters. Selenologists generally agree that the domes are of internal origin but the eruptive mechanism is disputed. Three hypothetical origins are: (1) shield complexes (extrusions of basic lava) like Mauna Loa, Hawaii, (2) laccoliths (intrusions of silicic lavas that arch up the surface) like the Green Mountains, Utah, and (3) pingoes (surfaces arched up by underlying ice or permafrost). Terrestrial shield complexes and laccoliths have sizes, shapes, and features analogous to lunar domes, but terrestrial pingoes are 10–100 times smaller, often have steeper slopes, and are geologically very short-lived. The existence of differences among the lunar domes suggests that there are both kinds of volcanic types (including composite ones), implying an advanced degree of differentiation. Caption: W. Cameron. Photo: NASA Langley Research Center.

PLATE 196B

Lunar Orbiter V Site 51, Frame 214, Medium Resolution (~50 meters) vertical view of some of the Marius domes, northwest of those seen in Plate 196A. Variations in dome morphology are seen that suggest a complex origin. The encircled domes appear to have steeper rising peaks, suggesting that later eruptions were more silicic and hence products of differentiation. Note also the abnormal clustering of craters on many of the domes compared to their surroundings. In the center are two sinuous rills, one of which starts in an elongated, crater-form structure. These channels are generally thought to be eroded by fluid material, such as water, lava (basic) or fluidized dust (silicic ash flows). Many terrestrial fluid lava channels, however, are usually lava tubes which generally have some of the roof remaining, often have vertical walls and overhangs, and are thus rarely continuous trenches. Lunar sinuous rills, to the contrary, are generally continuous, have wall slopes of about 30–40°, and are about 100–1000 times larger than the majority of terrestrial lava tubes. The two small rills in the upper right corner have sizes and characteristics, such as interruptions and steeper sides, more suggestive of lava tubes than the larger central continuous rills. The largest sinuous rill shown here is about 60 kilometers long, up to 2 kilometers wide, and about 100 meters deep near this source. Caption: W. Cameron. Photo: NASA Langley Research Center.

MARIUS

NORTH

5 km (horizontal)

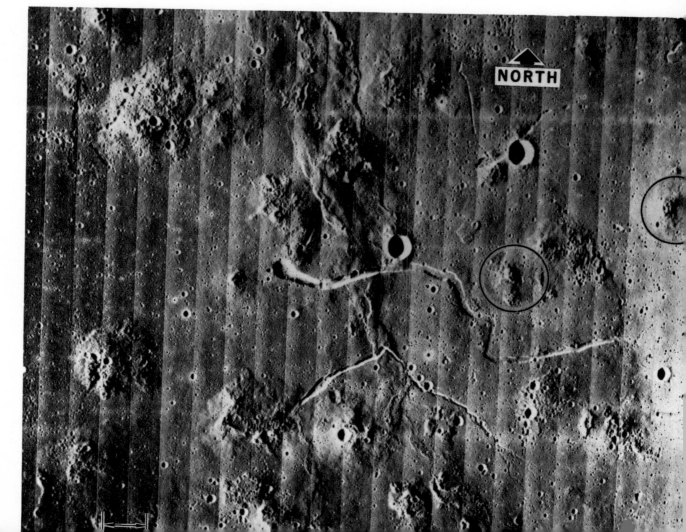

NORTH

PLATE 197A

Lunar Orbiter V, Site 48, Frame 199 (2nd of 3), High Resolution (∼3 meters) photograph of the northern half of the floor of the 40-kilometer diameter crater Aristarchus at 47° W, 23° N. (lunar coordinates) The central peak appears at the lower left and part of the northwest inner wall appears at the upper left corner. The distance along the bottom of the photo is about 19 kilometers. Note the very rugged floor with innumerable small hills and cracks or rills. Near the right edge, these cracks are parallel to the crater walls. Notice also, in the upper right, the peculiarly large ring of hills and domes. Observe as well the apparent flow markings on much of the floor material. The domical hills on the floor, some of which have elongate summit craters (e.g., left of center), and the obvious flow marks, suggest that volcanism has occurred. This crater and other areas in its vicinity are sources of repeated reports of transient phenomena, suggesting that mild activity or degassing is still taking place. Caption: W. Cameron. Photo: NASA Langley Research Center.

PLATE 197B

Lunar Orbiter V, Site 30, Frame 125 (2nd of 3), High Resolution (∼3 meters) photograph of the northwest portion of the floor of the 87-kilometer diameter crater Tycho at 11° W, 42° S (lunar coordinates), in the southern highlands of the Moon. At left is the lower part of the inner northwest wall of the crater. The distance along the bottom of the photograph is about 28 kilometers. Compare the floor seen here with that of Plate 197A above. Tycho is thought to be the youngest major feature on the visible side of the Moon (probably less than 100 million years old and possibly less than one million years old). There are very few subsequent craters, such as the one at the very bottom right center (∼1.2 kilometers in diameter; about the size of Meteor Crater, Arizona), to be found on or in it. The floor of Tycho is also characterized by properly ruggedness, with no smooth portions. In it, too, is evidence of flow in many places. One finds many of the hills to have odd-shaped, elongated summit craters, distinct from the characteristic circular craters which some assume to be impact and some assume to be volcanic. These suggest that some, if not all, of the domical hills are of internal origin. Caption: W. Cameron. Photo: NASA Langley Research Center.

PLATE 198A

Lunar Orbiter V, Site 46, Frame 189, Medium Resolution (∼30 meters) photograph of the region of the Prinz sinuous rills. Rills are usually designated by the nearest prominent crater in their vicinity. Prinz, the large crater (45-kilometer diameter) at the bottom, is located at 43° W, 27° N (lunar coordinates). This region contains the largest concentration of this comparatively rare type of rill (channel), whose characteristics are: (1) a crater at the higher, wider, deeper end, (2) narrower width and shallower depth as the channel descends, (3) arc lengths of small radius of curvature (∼1 kilometer), and (4) a course which is often influenced by topography. These characteristics are in contradistinction to the normal rills characterized by linearity and uninterrupted courses through all types of terrain. Note that all source craters are without external walls and several have irregular shapes. These are in contrast to most lunar craters, typified by the circular craters at the upper left, suggesting that they are different in kind and presumably in origin. The sinuous rills are generally thought to be channels eroded by passage of some fluidized material, such as: (1) water (or ice), (2) lava, or (3) ash flows, the latter two being volcanic in which the lava is basic (basaltic) and the ash flow is more acidic (andesitic to rhyolitic). See Plate 117A. Note also that several contain a smaller rill within the larger, e.g., the one that starts at the center of the outer rim of Prinz. Some appear to end in crater chains. This region also contains several domes, two of which occur between the three rills in the center of the photograph. The domes are generally thought to be of internal origin. This region is part of a larger one containing many features that suggest both mafic and felsic volcanism have occurred in this part of the Moon. Caption: W. Cameron. Photo: NASA Langley Research Center.

PLATE 198B

Lunar Orbiter V Site 23.1, Frame 95, Medium Resolution (∼30 meters) photograph of the Hyginus rill region at 6.5° E, 8° N (lunar coordinates). In the field of view are examples of several different types of features. The large crater (11 kilometers in diameter) in the center from which rills proceed in several directions is Hyginus. Near and from it emerge three types of rills. The eastern (right) branch of the main rill and some of the smaller, less distinct ones to the left are possibly grabens. The V-shaped rill trending southwest from Hyginus is the more frequent or normal rill type and may be a tension fault. The west branch of the main rill is the crater chain type. The craters in crater chains are generally thought to be of internal origin, possibly maars. Hyginus has no external wall and is irregular in shape. It, too, is thought to be of internal origin. Note the small mounds on its floor. Other evidence of volcanism in this region is found in the presence of a typical lunar dome with an elongated, eccentric summit crater (encircled) as well as apparent flow fronts or one edge only of ridges, seen at lower center to lower right and upper center. Caption: W. Cameron. Photo: NASA Langley Research Center.

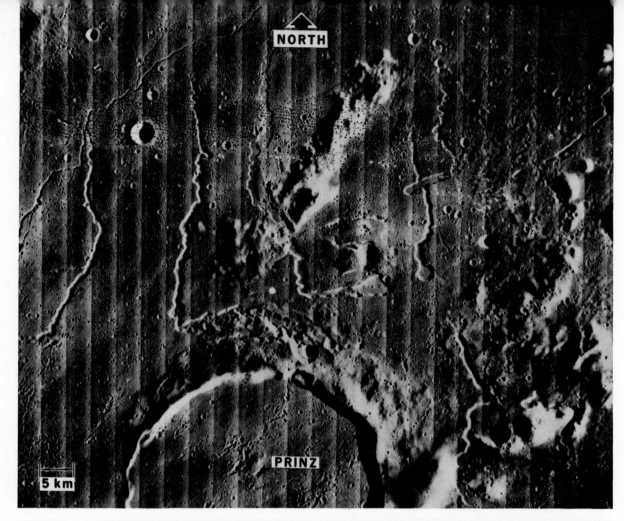

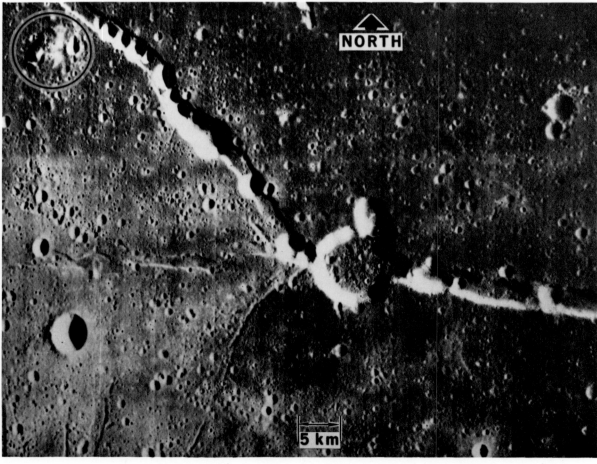

REFERENCES FOR THE
PLATE SECTION

Anderson, T. 1910. *Quart. Jour. Geol. Soc.* 66:631–633.

Anderson, R., *et al.* 1965. Electricity in Volcanic Clouds. *Science* 148:1179–1189. Aramaki, S. 1956. The 1783 Activity of Asama Volcano. *Jap. Jour. Geol. Geogr.* Part 1, 27:189–229.

Badgley, P. C. 1965. *Structural and Tectonic Principles.* New York: Harper and Row, 521pp.

Baker, P. E., *et al.* 1964. The Volcanological Report of the Royal Society Expedition to Tristan da Cunha. 1962. *Phil. Trans. of Royal Society*, London, Series A, 256,1075:439–578.

Banfield, A. F., Behre, C. H. and St. Clair, D. 1956. Geology of Isabela (Albemarle) Island, Archipielago de Colon (Galapagos). *Bull.Geol. Soc. Am.* 67:215–235.

Corcoran, R. E. 1953. *The Geology of the East Central Portion of the Mitchell Butte Quadrangie, Oregon.* Oregon Master's Thesis, 80pp.

Day, A. L. and Allen, E. J. 1925. The Volcanic Activity and Hot Springs of Lassen Peak, California. *Pub. 360 Carnegie Inst.* Washington, 190pp.

Deer, W. A. 1969. Field Excursion Guide to the Tertiary Volcanic Rocks of Ardnamuchan. *Intern. Assoc. of Volc. and Chem. of Earth's Interior Symposium,* Oxford, 29pp.

Denny, C. S. *et al.*, 1968. A Descriptive Catalog of Selected Aerial Photographs of Geologic Features in the United States. *U.S. Geol. Surv. Bull. 965-D.*

Elston, W. E. 1957. Geology and Mineral Resources of Dwyer Quadrangle, Grant, Luna and Sierra Counties, New Mexico. *New Mexico Bur. Mines Bull. 38*, 86pp.

Foshag, W. G. and Jenaro Gonzolez, R. 1956. Birth and Development of Parícutin Volcano, Mexico. *U.S. Geol. Surv. Bull. 965-D.*

Foster, H. and Mason, A. C. 1955. 1950 and 1951 Eruptions of Mihara-Yama O'Shima Volcano, Japan. *Bull. Geol. Soc. Am.* 66:731–732.

Friedman, J. D. 1968. Thermal Anomalies and Geologic Features of the Mono Lake Area, California, as Revealed by Infra-Red Imagery. *Earth Resources Aircraft Program Status Review.* Houston: NASA Manned Spacecraft Center, pp. 11–1 to 11–76

Friedman, J. D., Williams, R. S. Jr., Pálmdson, G. and Miller, C. O. 1969. Infrared surveys in Iceland-Preliminary Report. U.S. Geol. Sur. Prof. Paper 650-C, p. C-89–C-105.

Green, J. 1965. *A Study of the Feasibility of Using Nuclear Versus Solar Power in Water Extraction from Rocks.* Air Force Cambridge Research Lab. Final Report, 353pp.

Guest, J. E. 1969. Upper Tertiary Ignimbrites in the Andean Cordillera of part of the Antofagasta Province, Northern Chile. *Bull. Geol. Soc. Am.* 80:337–62.

Heinrich, E. W. 1966. *The Geology of Carbonatites.* New York: Rand McNally and Co, 555pp.

Hamilton, W. M. and Baumgard, I. L. 1959. White Island. *New Zealand Dept. Sci. Indust. Res. Bull. 12T*, 84pp.

Hartmann, W. K. 1967. Secondary Volcanic Impact Craters at Kapoho, Hawaii and Comparisions with the Lunar Surfaces. *Icarus* 7:66–75.

Huber, N. K. and Rinehart, C. C. 1965. *Geologic Map of the Devil's Postpile Quardrangle, Sierra Nevada, California.* U.S. Geol. Surv. Quad. Maps of U.S. GQ-437, scale 1:62500.

———— 1966. *Cenozoic Volcanic Rocks of the Devil's Postpile Quadrangle, Eastern Sierra Nevada.* U.S. Geol. Surv. Quad. Maps of U.S. GQ-437, scale 1:62500.

Lacroix, A. 1904. Montagne Pelée et ses Eruptions. Paris: Masson et Cie, 662pp.

Loomis, B. F. 1958. *Lassen Volcano, A Pictorial History,* revised ed. Mineral, California: Loomis Museum Assoc., Lassen Volc. Nat. Park.

Lovering, T. S. 1957. Halogen-Acid Alteration of Ash at Fumarole No. 1, Valley of Ten Thousand Smokes, Alaska. *Bull. Geol. Soc. Am.* 58:1585–1604.

MacDonald, G. A. and Orr, J. B. 1950. The 1949 Summit Eruption of Mauna Loa, Hawaii. *U.S. Geol. Surv. Bull.* 974-A:1–31.

———— and Eaton, J. P., 1957, Hawaiian volcanoes during 1954. U.S. Geol. Survey Bull., 1061-B, pp. 17–72.

————— and Powers, H. A. 1968. A Further Contribution to the Petrology of Haleakala Volcano, Hawaii. *Bull. Geol. Soc. Am.* 79:877–888.

Minakami, T. and Peck, D. L. 1968. The Formation of Columnar Joints in the upper Parts of Kilauean Lava Lakes, Hawaii. *Bull. Geol. Soc. Am.* 79:1151–1166.

Osgood, J. and Green J. 1965. Sonic Velocity and Penetrability in Simulated Lunar Materials. *Geophysics* 30:536–61.

Perret, F. A. 1937. The Eruption of Mt. Pelée. *Pub. 458, Carnegie Inst.* Washington, 126pp.

Peterson, N. V. and Groh, E. A. 1964. Diamond Craters, Oregon. *The Ore Bin* 26:17–36.

Reck, H. 1915. *Physiographische Studie über Vulkanische Bomben.* Erganzungsband, Zur Zeitschrift für Vulkanologie 1914–15, Herausgeber: I. Friedlaender, Neapel, D. Reimer (E. Vohsen) Berlin, 224pp.

Richards, A. F. 1960. *Rates of Marine Erosion of Tephra and Lava at Isla San Benedicto,* Mexico. Section 10, Copenhagen: 21st Int. Geol. Congress.

Rittmann, A. 1962. *Volcanoes and Their Activity.* New York: John Wiley, 305pp.

Robinson, C. L. 1956. Geology of Devil's Tower, Wyoming. *U.S. Geol. Surv. Bull.* 1021-I:289–302.

Smith, R. L. and Baily, R. A. 1968. Resurgent Cauldrons, In *Studies in Volcanology*, R. R. Coats, *et al. Geol. Surv. Am. Memoir 116*: 613–662.

Symons, G. J. 1888. *The Eruption of Krakatau and Subsequent Phenomena.* London: Report of Krakatau Comm. of Royal Soc.

Thorarinsson, S. 1964. *Surtsey, Almenna Bókáfélagid, Iceland.* 64pp.

Van Bemmelen, R. W. and Rutten, M. G. 1955. *Tablemountains of Northern Iceland.* (and related geological notes) Leiden: E. J. Brill, 217pp.

White, D. E. 1957. Thermal Waters of Volcanic Origin. *Bull. Geol. Soc. Am.* 68:1637–1658.

Williams, H. 1932. Geology of Lassen Volcanic National Park, California. *Bull. Dept. Geol. Sci.* 21.

————— 1936. Pliocene Volcanoes of Navajo-Hopi Country. *Bull Geol. Soc. Am.* 47:111–172.

GLOSSARY

An asterisk [*] suggests that the term be eliminated in lieu of a common synonym or descriptive adjective.

A dagger [†] suggests the better or best definition.

Material in brackets was added to the original by one of us [J.G.] for clarity or for expansion of the meaning to include extraterrestrial features or phenomena.

The original unedited material was supplied courtesy of the American Geological Institute.

aa

† Stokes and Varnes (1955), p. 1: "The Hawaiian word for solidified lava characterized by an exceedingly rough, jagged, or spinose surface. Most of the surface is covered with a layer or loose fragmental clinkery material a few centimeters thick. Below the surface the clinker fragments may be stuck firmly together. Typically, the upper clinker layer is underlain by, and grades into, a central massive layer (after G. Macdonald)."
Synonym: aphrolithic lava

Johannsen (1939), p. 163: "The native Hawaiian word for rough, scoriaceous lava, consisting chiefly of sharp, angular fragments of compact lava and rough clinkers cemented together." (Macdonald, G., in Hess, and Poldervaart, 1967); "Hawaiian word, introduced by C. E. Dutton in 1884, Aa is characterized by "rough, jagged, spinose, and 'clinkery' surface" —p. 15: Often called "scoriaceous", but "clinkery" is a better term; most conspicuous feature is "exceedingly rough fragmental top."
Eruptions of continental plateau regions and oceanic shield volcanoes are "very largely typical aa."

ac accessory ejecta
Calkins, F. (personal communication): "Ap-

plied to pyroclastic materials [ejecta] derived from previously solidified volcanic rocks of consanguineous origin; e.g., the debris of earlier lavas and pyroclastic rocks from the same cone. Such ejecta correspond to the 'matériaux paleogénes' of Lacroix and resurgent authigene products of von Wolff (1914)."
Cf. "accidental" [After Wentworth and Williams, (1932, p. 45.)]

accidental ejecta
Macdonald, G., in Hess and Poldervaart (1967), vol. 1, p. 51: fragments of nonvolcanic rock or volcanic rock totally unrelated to the eruption that ejected them.
Lacroix term: "matériaux énallogénes."

accidental tuff
Nelson and Nelson (1967), p. 2: "A tuff or pyroclastic rock composed of fragments and debris ejected from a choked-up volcanic vent or torn from the surrounding rocks."

accretionary lapilli
Macdonald, G., in Hess and Poldervaart (1967), vol. 1, p. 50: Wentworth and Williams (1932), p. 37: "[Term also known as volcanic pisolites] spheroidal masses of weakly to well-cemented ash, generally between 2 and 10 mm, but occasionally up to 30 or possibly even 60 mm, in diameter"—p. 51: Most commonly formed "by accretion of particles around a wet nucleus falling through a cloud of ash. The resulting 'mud balls' flatten somewhat on striking the ground. Some of them may roll on the surface of loose ash and grow like snowballs rolling down hill, thus acquiring a spiral structure."

accretionary lava ball
Macdonald, G., in Hess and Poldervaart (1967), vol. 1, p. 16: "Rounded mass formed by rolling up of viscous lava around some solid core, [e.g., fragment of partly or wholly consolidated bank

of the lava river; often occurs on aa flow tops and ranges in diameter from few cm to 5 m or more."

acid fumarole
Schieferdecker (1969, p. 275, term 4481): "name given by de Lapparent (1906) for fumaroles with temperatures between 200–800 degrees C., the escaping gases being HCl, SO_2, NH_3, Cl, H_2O."
Cf. "alkalic fumarole."

acid lava
Monkhouse (1965), p. 3: "A mass of molten igneous material, flowing slowly from a volcanic vent, stiff and viscous, rich in silica, and with a high melting-point [about 850°C,]. Hence, it solidifies (p. 4) rapidly and does not flow far, forming a steep-sided dome; e.g., Mount Lassen, Cascades, California. Some acid lavas solidify in a fine-crystalline state as rhyolite or dacite, others in a glassy form as obsidian."
Cf. "basic lava."

acid magma
Bayly (1968), p. 34: "preferred term is silicic or felsic, Magmas constituting by weight:

SiO_2	65–75
Al_2O_3	12–16
Fe_2O_3 plus FeO	4–8
MgO plus CaO	4–6
Na_2O plus K_2O	6–9."

acid solfatara
Von Wolff (1914): "A solfatara 'containing SO_3, SO_2, H_2O and CO_2'."

active volcano
Stamp (1961), p. 6: "A volcano from which periodical eruptions still occur." Contrast with "dormant" and "extinct."

ad **adventive cone**
Schieferdecker (1959), p. 239, term 4003: "A tuff or lava cone on the flank or foot of the major volcano, often on a radial crack as at Mt. Etna."
Less-preferred synonym: "parasitic cone,"
                "lateral cone,"
                "minor cone" (no q.v.),
                "subsidiary cone" (no q.v.).

adventive crater

Schieferdecker (1959), p. 241, term 4024: "A small crater developed on the flank or at the foot of a major volcano."
Less-preferred synonym: "parasitic crater,"
                "lateral crater,"
                "subcrater."

adventive neck
Schieferdecker (1959), p. 235, term 3939: "A vent not belonging to the main crater, but to an adventive crater of the volcano."
Less-preferred synonym: "subsidiary neck" (no q.v.).

adventive vent
Schieferdecker (1959), term 4053: "A chimmey emanating from satellitic injections such as small laccoliths and sills (Daly, 1914) in a parasitic volcano."
Less-preferred synonym: "peripheral vent" or "subordinate vent."

ag **agglomerate**
Stokes and Varnes (1955), p. 2: "A [non-stratiform] mass of unsorted volcanic fragments which may be loose or consolidated into a solid mass by finer volcanic material which fills the interstices. It is usually localized within volcanic vents or [within a few kilometers of them]. Some writers use the term as synonymous with 'explosion' or 'volcanic breccia" and maintain that the fragments should be of angular shape; others conceive the fragments to be mostly rounded. In any event the shapes of the fragments are not determined by action of running water but are the result of [eruptive activity]."

agglomerate lava
Shieferdecker (1959), p. 261, term 4304: "A lava which picked up bombs, slag and ash, or the loose blocks on the surface of a block lava stream, during its flow."

agglutinate [agglutinite-obsolete]
Macdonald, G., in Hess and Poldervaart (1967), vol. 1, p. 47: "Welded spatter" in which the "outsides of the fragments are still sufficiently fluid to stick together."

agglutinate cone
Macdonald, G., in Hess and Poldervaart (1967): "Same as 'spatter cone'."

ai **air-fall deposition**
[Pyroclastics accumulating by falling through

air. Vacuum-fall deposition would be the equivalent on volcanic cosmic bodies devoid of an atmosphere.]

* **air volcano**
Webster 3 (1967), p. 47: "A vent from which large volumes of volcanic or non-volcanic gases are discharged along with mud and stones."

al **alkalic fumarole**
Schieferdecker (1959), p. 275, term 4482: "Name given by de Lapparent (1906) for fumaroles emitting NH_4Cl, H_2S, CO_2, and H_2O at temperatures of 100–200 degrees C."

alloclastic
Term introduced by Wright and Bowes, (1963), pp. 79–86. "Volcanic breccias composed of fragments alien to the petrography of the volcano."

almond-shaped bomb
Schieferdecker (1959), p. 257, term 4244: "Spindle-shaped bomb."

am **amygdale**
Hess after Broderick, (1936), p. 518: "Gas cavities or vesicles in lava which have been filled by chalcedony or other minerals."

amygdule
[Diminutive form of amygdale.]

ap **apalhraun**
Friedman, J. D., (personal communication): "Icelandic term for block lava and aa lava."

* **aphrolithic lava or aphrolite**
Schieferdecker (1959), p. 249, term 4133: "Aa lava."

* **aphrolith**
Holmes (1928), p. 33: "A term, meaning 'foam-stone' applied to block lava or aa lava."
Jaggar (1917).

aq **aquagene tuff**
Fisher (1966), in *ESR* vol. 1, p. 292: Carlisle, (1963), p. 48–71 [Tuff deposited underwater].

ar **areal eruption**

Schieferdecker (1959), p. 236, term 3953:
1) "As understood by von Wolff (1929) it is Daly's hypothesis (1929) of the deroofing of a batholith. Overflow of magma, welling up from the deroofed area might produce lava floods around its periphery."
2) "As understood by Reck it is a hypothesis according to which plateau basalt poured forth from clusters of vents in central areas instead of from fissures."
† 3) [Simultaneous volcanic activity from associated vents encompassing a large but descrete region.]

as **ash**
[Usually angular fragments 0.05–4 mm in diameter. If volcanic they are derived by explosion [not burning] and constitute one class of pyroclastics.]

ash tuff
Schieferdecker (1959), p. 258, term 4265: "Mainly consisting of volcanic ash, mixed with only minor coarser material." Blyth (1940) distinguished: "coarse ash tuff [0.5–4 mm] and fine ash tuff [0.05–0.5 mm]."

ash avalanche (ash flow)
Schieferdecker (1959), p. 273, term 4453: "[Hot] ash avalanche"—"fresh-fallen hot sand and ash, descending the flank of a volcano with the velocity and aspect of a snow avalanche, but generally in an impressive silence. The accumulation of ejected material upon the upper portions of the cone, consisting principally of fine intensely hot ash, in fact is in an unstable equilibrium and a peculiar state of mobility (Perret, 1924)."
Less-preferred synonym: "ash slide."

ash bed
Snyder, J., (personal communication): "Bed of rock particles of volcanic origin that were deposited in a solid or semi-solid state. Grades into true sedimentary rock if volcanic material has been reworked by streams, wind, etc. May also be gradational into true flows through ignimbrites or welded tuffs."

ash cloud
Schieferdecker (1959), p. 269, term 4411: "Eruption cloud consisting of volcanic ash and gases."
Less-preferred synonym: "dust cloud."

ash cone

Monkhouse (1965), p. 21: "A small volcanic cone of ash; its shape depends on the nature of the material, but is usually concave due to the spreading outwards of material near the base, and is less steep than a 'cinder cone'." Synonym: "tuff cone."

ash fall
Ross and Smith (1961), p. 3: "Deposition of volcanic ash directly from the air, [or from altitude,] generally, but not always, resulting in a stratified deposit showing crude to very complete sorting of its component parts." Synonym: "ash shower."

ash flow
† Ross and Smith (1961), p. 3: "A turbulent mixture of gas and pyroclastic materials of high temperature, ejected explosively from a crater or fissure, that travels swiftly down the slopes of a volcano or along the ground surface. The solid material in an ash flow although unsorted, is dominantly of particles of ash size [less than 4 mm in diameter] but generally contains different amounts of lapilli and blocks." Synonym: "pyroclastic flow."
Cook (1966), p. 159: "Lower part of nuée ardente or gas "cloud" consisting of a incandescent density or turbidity current, carrying the greater part of the solid and liquid matter of the "cloud," and from which rises a continuously dissociating cloud of gas and fine particles."

ash-flow tuff
Ross and Smith (1961), p. 3: "The consolidated deposits of volcanic ash resulting from an ash flow are called ash-flow tuff. Ash flow is here used as an adjective to indicate the mechanism of dispersal, and tuff indicates the state and size of the material. Ash-flow tuff is an inclusive, general term for consolidated ash-flow beds that may or may not be either completely or partly welded."

ash field
Webster 3 (1967), p. 128: "A thick widespread deposit of volcanic ash, called also ash plain."

ash layer
Schieferdecker (1959), p. 255, term 4217: "A deposit of the finest volcanic material. French term is 'lit de cendre' or 'couche de cendre'." [If reworked by wind or water the term can be applied to sedimentary rocks also.]

ash plain

See also "ash field."

ash rock
Webster 3 (1967), p. 128: "Rock consisting of volcanic ash."

ash shower
See also "ash fall."

ash slide
Schieferdecker (1959), p. 273, term 4453: "Ash avalanche."

as   ash spread
*   Webster 3 (1967), p. 128: "Ash shower."

*   ashstone
"Indurated volcanic ash composed of pyroclastic rock from [0.05–4 mm] in diameter." Hess [after Kotô (1916), p. 13]. "An ash-gray, friable, porous hypersthene-trachyandesite which looks like hardened volcanic ash; used for building material in Japan because it is easily quarried."

*   asperite
Brown and Runner (1939), p. 27: "A collective name for the rough cellular lavas whose chief feldspar is plagioclase."

at   ataxite
Holmes (1928), p. 38: Loewinson-Lessing, 1911: "A brecciated or irregularly mottled composite volcanic rock, in which the broken fragments of one lava-flow are irregularly distributed in another. A similar structure to which the term may also be applied occurs in certain minor intrusions."

athrogenic
Webster 3 (1967), p. 138: "Of or relating to clastic rocks of igneous origin."

atrio
Webster 3 (1967), p. 139: "The valley between two cones or a volcano."

†   Schieferdecker (1959), p. 241, term 4023: "The generally crescent-shaped valley situated between the central cone and the somma ring. Called atrio after a similar valley: 'atrio del cavallo' of Vesuvius." Less-preferred synonym: "fosse."

au   authigenous ejecta

Schieferdecker (1959), p. 255, term 4207: "[Ejecta chemically and petrologically similar to the "normal" ejecta of the vent.]"

au   Challinor (1964), p. 31: "A volcanic breccia— a lava broken up during cooling, either by the escape of steam, etc., or by the hardening and breaking-up of the surface in the process of flow [the latter specified still further as 'flow-breccia']."

autobrecciation
[The process of forming and cementing volcanic fragments (>4mm dia.) resulting from breakage of a cooling lava surface by, 1. escape of gases, or 2. flow processes. In the latter case, the breccia is termed a "flow-breccia" (no. q.v.).]

ax   axiolitic
†   [Pertaining to a type of microspherulitic growth.]
Ross and Smith (1961), p. 4, from Iddings (1899) p. 419: "In certain kinds of rhyolite, apparently composed of welded glass fragments, there is a microspherulitic growth which bears a definite relation to the form of the supposed glass fragments. The feldspar fibers are in groups which are approximately normal to the outline of the fragments and radiate inward."

ba   barnacle stalactite
Jagger T. A., (1931), Lava stalactites, toes, and squeeze-ups, Volcano Letter, 345, p. 1–3: "An irregular excrescence of molten lava that has oozed into spiracles through cracks in the chilled wall."

barranco
Lyell (1871), Principles of Geology. "They are very shallow ravines near the summit, but become rapidly deeper and have precipitous sides towards their terminations. . . ." [A barranco can evolve by erosion or tectonics to become an entrance valley draining a volcano or caldera.]
Cottingham (1951), p. 162: "Radial corrugation [or deep valley] on a volcanic cone resulting from mudflows and their erosive effects caused by the steep gradient."
†   [A deep valley which may or may not be associated with volcanics.]

basal scoria
["Scoria composing the solidified base of a lava flow, over-ridden by the overlying moving lava."]
Less-preferred synonym: "ground moraine,"
                        "basal moraine,"
                        "basal clinker,"
                        "scoria moraine."

See also: "basalt" or "basaltic dome."

basalt globes
Schieferdecker (1959), p. 262, term 4311: "[Friable but] quite dense or slightly porous lava nodules with thick tachylitic [sideromelane] cooling rims, fracturing along radial and tangential planes, (van Bemmelen and Rutten, 1955)."
Less-preferred synonym: (no q.v.) "basaltic lava nodules."

basaltic
Webster 3 (1967), p. 180: "Relating to, formed of, containing, or resembling basalt."

basaltic cinder cone
See: "cinder cone."

basaltic plateau
Cotton (1944), [A type of basalt plateau]; "basalt plain."

*   basaltine
"Basaltic."

basic lava
Monkhouse (1965), p. 32: "A mass of molten igneous material on the surface of the earth, rich in iron, magnesium and other metallic elements, and relatively poor in silica. It has a low melting-point and will flow readily for a considerable distance before solidifying. Usually its flow from the vent of a volcano is unchecked and free from much explosive activity. The basic lava may form a large cone [a shield volcano], or, if it flows from number of fissures, an extensive plateau."
Cf. "acid lava."
Recommended synonym: "mafic lava."

basic magma
Bayly (1968), p. 34: percent by weight:

SiO_2	48–58
Al_2O_3	13–17
Fe_2O_3 plus FeO	9–14
MgO	5–8
CaO	8–12
Na_2O plus K_2O	3–5

Recommended synonym: "mafic magma."

base surge
[An outwardly moving ring of debris emanating from an explosion center and traveling along the surface of the ground.]

be   **bed**
Monkhouse (1965), p. 36: "[In volcanology] applied to a layer of pyroclasts, [e.g., an ash-bed] and to [unit flows] in a sequence of lava flows."

bedded deposits
[In volcanology, volcanic material, often alternating with lava flows or tuff units.]

bedding voids
[Interflow voids produced when one flow covers another or when one flow advanced over its own shattered crust.]

bedded
Stokes and Varnes (1955), p. 14: "A volcano whose cone consists of layers of tuff and lava."

belt of fire
Webster 3 (1967), p. 202: "An area of active volcanoes." [Often used to describe the volcanologically active region that encircles the Pacific Basin.]

bench lava
[Old, already frozen lava in craters forming platforms; in distinction to the active magma in the crater lake or the crater floor.]

bl   **blister**
Webster 3 (1967), p. 235: "An elevated layer of rock resulting from the flow of molten rock into low wet areas and the generation of steam pickets."
†   Macdonald, G., Hess and Poldervaart, (1967), vol. 1, p. 13: "Blisters form on skin of lava flow by gas puffing up beneath flow; usually less than 1 m across and 70 cm high [in Hawaii], although some are observed 3 m in diameter." See also "lava blister" and "magma blister."

blister cone
Webster 3 (1967), p. 235: "A small cone produced by the expansion and escape of gas or vapor from liquid lava."

block
Ross and Smith (1961), p. 4: Blyth (1940,

p. 147), "[In volcanology,] a fragment [>3.2 cm. dia.] of cognate or accidental material larger than lapilli and usually angular, which has been erupted in a solid or nearly solid state."

block flow
Ross and Smith (1961), p. 7: "Flow characterized by coarser material than in pumice flow." See also "lapilli flow" and "blocks."

block lava
Stamp (1961), p. 67: quotes Cotton (1944), p. 153: "As used by many authors 'block lava' is synonymous with 'aa'."
Finch (1953), pp. 746–770: A contrast exists between the typically spinous "clinkers" formed of a true aa surface and the blocks of angular form of which some lava fields, notably those consisting of the more acid lavas, are superficially composed. Finch suggests that the descriptive term "block lava" should be reserved for these latter.

blowhole
Schieferdecker (1959), p. 243, term 4055: "A miniature crater on the surface of a thick lava flow, very often evolving into a driblet cone (Daly, 1914). Kemmerlin (1922) and Taverne (1926) used the term for small maars or for boccas."
†   [Any small craterlet produced by rapid to near explosive gas discharge.]

*   **blowing cone**
Webster 3 (1967), p. 237: "A small volcanic cone built up of congealed drops of lava from which steam or other vapors escape."
See "driblet cone."

blowpipe flame
Schieferdecker (1959), p. 266, term 4375: "[Volcanic] flames with a temperature up to 1300°C, as e.g., occurring at gas vents of Kilauea."

bo   **bocca**
Webster 3 (1967), p. 245: "[In volcanology] a vent on the side or near the base of an active volcano from which lava issues." [Obvious adjectives can be applied, i.e., explosion, gas, lava, adventive, basal.]

boiling mud pit
[In fumarolic areas, a craterlet filled with a hot, gas-charged hydrothermal clay-water-steam

mixture. The mud is not necessarily at or above its boiling point.]

boiling spring
Webster 3 (1967), p. 247: "A natural pool of hot water through which bubbles of steam or volcanic gas rise to the surface often with much force." [The water is not necessarily at or above its boiling point.]

bomb
Stokes and Varnes (1955), p. 16: "[In volcanology] a mass of lava [over 3.2 cm in diameter] projected by explosion from a volcanic vent. Bombs are viscous when expelled but usually cool and assume [ellipsoidal, discoidal, or irregularly rounded] forms and characteristic surface markings before striking the earth." [Fragments smaller than 3.2 cm but over 4 mm are called lapilli. The following shape, textural or compositional adjectives are applied to bombs: "rotational bomb," "tear-shaped bomb," "spheroidal bomb," "spindle-shaped bomb," "pancake-shaped bomb," "ribbon bomb," "cored bomb," "olivine bomb," "breadcrust bomb," "explosive bomb," "cow-dung bomb," "slag bomb," "pumiceous bomb." Bombs are characterized by a well-defined crust and may be internally cellular or hollow.]

bomb pit
Schieferdecker (1959), p. 265, term 4362: "A small craterlet formed by the impact of volcanic bombs or blocks falling in sandy or ashy deposits on the slope or foot of the volcano [in the vicinity of a volcanic vent]."

bomb sag
[Depressed and/or deranged stratified material around a volcanic bomb or block produced early by the impact or weight of the bomb on the deposit or later by differential compaction effects around it.]

bore
Webster 3 (1967), p. 255: "A [tubular] surface opening or outlet [as of a geyser]."

br   breached cone—breached crater
[A crater or cone, one portion of the wall of which is destroyed by erosion, by water or lava or a directed volcanic blast. Water or lava erosion may be localized by a structural zone of weakness, a flank eruption, or overflow at a low point on the cone rim. In rare cases a cone or

crater may be breached on the side facing a body of water by wave or sapping action. Some horseshoe-shaped craters may be caused by a crescentic deposition of pyroclastics and are not the same as breached craters.]

breccia pipe
[A tubular vent as of a diatreme or volcanic neck, filled with angular (and also sometimes rounded) fragments. The breccia may exist in concentric zones.]
Synonym: "vent breccia."

breadcrust bomb
Macdonald, G., in Hess and Poldervaart, (1967), vol. 1, p. 49: "A bomb formed by continued vesiculation and swelling of the core of the bomb after the outer skin has become too rigid to stretch. The result is a network of surface cracks, many with nearly vertical sides and wide bottoms, resembling those on a loaf of thick-crusted bread."

breadcrust structure
[The surface texture of breadcrust bombs q.v.]

bu   bulbous dome
H. T. Stearns, (personal communication): "A dome-shaped mass over a vent, formed by outpouring of viscous lava."

*   burning mountain
Webster 3 (1967), p. 300: "Volcano."

ca   caldera
"A large volcanic depression, more or less circular or cirquelike in form the diameter of which is many times that of the included vents." (Williams, 1941). [A polygenetic, more or less circular volcanic depression over 1 km and less than 250 km in diameter formed by subsidence or engulfment with or without varying degrees of explosivity and localized by circular or polygonal fractures. Vents may occur on the floor, rim, or flank. Structures greater or less than 1 km may be termed maars, volcano summit pits, or sinks and over or under 250 km, volcanic basins or volcano-tectonic depressions.]

caldera complex
[A group of associated calderas within a discrete province as in the Tibesti mountains.]

caldera island
Schieferdecker, (1959), p. 238, term 3982: "[A

ring or partially ring-shaped] island of which only the caldera [wall] rises above the water."

caldera lake
[A lake occupying or partially occupying the floor of a caldera.]

caldera rim
[The crest of the caldera wall along any radius profile.]

caldera ring
[The perimeter or partial perimeter walls of a caldera.]

*   capillary lava
[Pele's hair.]

carapace
Schieferdecker (1959), p. 242, term 4040: "The frozen surface of a lava dome" [or any lava surface]; "also, the crust of a tholoid" (A. Poldervaart).

cascade
[A lava flow or pattern produced by overflowing or jetting mechanisms.]

cauldron
[A structure exposing the lower level of a caldera, i.e., ring dikes, volcanic necks, ignimbrite remnants, etc.]

cauldron-subsidence
Holmes (1928), p. 55: "The sinking of part of the roof of an intrusion within a closed system of peripheral faults up which magmas have penetrated."

cavernous
Johannsen (1939), p. 204, "Porous texture not due to expansion of gases, but to the weathering out of certain constituents. If small and regularly cavities occur, the texture is said to be porous, if larger and with corroded walls and irregular distribution, cellular."

ce   central depression
Strahler (1963), p. 603: "[A sink] produced by a subsidence that follows withdrawal of magma from below; a pit crater on the floor of a depression."

central cone
Schieferdecker (1959), p. 240, term 4010: "A cone [more or less centrally located] within the main crater in a volcano [summit pit] or in a caldera. The small cone as was found at Vesuvius in October 1904 was called 'terminal conelet' by Perret (1924) and 'volcanello' by Italian [volcanologists]."

central eruption
Monkhouse (1965), p. 58: "An eruptive form of volcanic activity which proceeds from a single vent or a group of closely related vents, in contrast to a linear or 'fissure eruption'. The product of a central eruption is a cone of some kind."

central vent
Stokes and Judson (1968), p. 498: "An opening in the earth's crust, roughly circular, from which magmatic products are extruded. A volcano is an accumulation of material around a central vent."

central volcano
[A volcano more or less centrally located within a caldera. Do not confuse with volcano of the central eruption type which is a volcano resulting from a continued eruption from one site.]

ch   chalazoidite
"An ellipsoidal [to spheroidal] body composed of [pyroclastic] fragments ranging in size from small grains to a half-inch in diameter, most of them have a nucleus of looser texture, around which are concentric layers of alternating density and an outside layer of greater strength. (Hess after Berry, 1929, p. 130)."
Synonym: "volcanic pisolite, " "volcanic mud ball," "volcanic muddrop."

chaotic tuff
Schieferdecker (1959), p. 259, term 4275: "Massive, non-stratified tuff consisting of a mixture of fine and coarse material equally distributed, as in nuée ardente deposits (A. Rittmann)."

chlorine fumarole
Schieferdecker (1959), p. 275, term 4479: "Name given by von Wolff (1914) for a fumarole emitting chlorine gas [also HCl, SO_2, CO_2 and H_2O] at temperatures above 800 degrees C."

chilling effect
Schieferdecker (1959), p. 270, term 4426: "The cooling influence of already frozen lava on pyromagma, causing among other effects a

temperature drop of the gaseous emanations as in a lava lake (Perret, 1924)."

chimney
Webster 3 (1967), p. 389: "A pipelike more or less vertical natural vent or opening [as for example] the conduit of a volcano."

chrysophyric
H. Hess, U.S. Bureau of Mines (1968). "A basalt with phenocrysts of olivine."

ci   cinder
Stokes and Varnes (1955), p. 23: Uncemented volcanic fragment, produced during an eruption ranging from 4 to 32 mm diameter—usually glassy or vesicular [also called lapilli], produced when lava is highly charged with gas."

cinder cone
Macdonald, G., in Hess and Poldervaart, (1937), vol. 1, p. 53: "Cone formed by explosive eruptions or lava fountains, composed almost wholly of loose material, predominantly cinder; also ash and spatter but it ranges in height up to several hundred meters and in basal diameter up to several km. The cone may be beautifully symmetrical [with slopes between 30–40°] and a nearly circular ground plan. The vent is commonly cylindrical."

cl   clastolithic accumulation
Schieferdecker (1959), p. 228, term 3839: "Heavier, semi-solid, pasty epimagma sinking through the frothy pyromagma (Jaggar)."

clastogenetic lava
Schieferdecker (1959), p. 262, term 4307: "A veneer of welded lava "rags" without ash, resulting from lava fountains (A. Rittmann)." Less-preferred synonym: "pipernoidal lava."

clastogenetic lava flow
Schieferdecker (1959), p. 262, term 4306: "Flows originating from lava fountains which accumulate liquid lava spatter to such a degree, that they weld together and flow downhill as a compact mass (A. Rittmann)." Less preferred synonym: "rootless lava flow."

cleaved volcano
Schieferdecker (1959), p. 247, term 4105: "Sapper (1927), a second stage of volcano destruction by erosion, deep and wide valleys having destroyed the regular form of the vol-

cano and its crater rim." Synonym: "dissected volcano," "furrowed volcano," "volcanic wreck."

clinker
Macdonald, G. in Hess and Poldervaart, (1967), vol. 1, p. 15: "In Hawaii, geologists refer to "fragmental portions of aa flows" as "clinker" because of the resemblance to clinker formed in the grate of a furnace; most clinker fragments are less than 15 cm across."

clinker breccia
Schieferdecker (1959), p. 261, term 4293: "Flow breccia"

clinker field
Schieferdecker (1959), p. 249, term 4136: "A large area covered with block lava or aa lava." Preferred synonym: "aa field."

co   cognate
Webster 3 (1967), p. 440: "[In volcanology,] pertaining to volcanic fragments in solidified lava which are part of the same extrusion" [or inclusions or breccia fragments which are petrographically similar to the matrix]."

cognate ejecta
Schieferdecker (1959), p. 255, term 4206: "Tyrrell (1928), material ejected directly derived from or related to the vent magma (Blyth, 1940)"

cold fumarole
Schieferdecker (1959), p. 275, term 4484: "Name given by de Lapparent (1906) for fumaroles emitting H_2O with or without CO_2 at temperatures below 100 degrees C."

cold mudflow
Schieferdecker (1959), p. 271, term 4438: "A flow which originates when the already cooled ash and debris cover on the volcano slope is washed down by heavy rains [see rain lahar]. It may also originate by the collapse of the wall of a crater lake or by melting ice caps." Synonym: "cold lahar."

collapse
[In volcanology, a subsidence resulting in the formation of a basin.]

collapse caldera
[A caldera formed primarily by subsidence.] Synonym: "sunken caldera."

collapse canal
Schieferdecker (1959), p. 251, term 4166: "[A linear depression] in the surface of a flow after the roof of an empty lava tunnel collapsed."

composite cone
Nelson (1967), p. 82: "A volcanic cone formed when lavas and fragmental material are erupted alternately; applicable to most large cones."

composite neck
Schieferdecker (1959), p. 244, term 4062: "[A vent filling] composed of several intrusions of different magmas."

composite volcanic cone
Monkhouse (1965), p. 72: "A cone built up over a long period of time as the result of a number of eruptions, consisting of layers of ash, cinder and lava fed from the main pipe, which culminates in a crater. It is often known in U.S.A. as a 'stratovolcano'." Examples include Mt. Hood (Oregon), Etna (Italy), and Fuji-yama (Japan).

composite volcano
[A stratovolcano], compare "composite cone."

compound crater
Schieferdecker (1959), p. 240, term 4014: "[An intersecting group of craters] resulting from the shifting of the explosion center within the [crater group area]. Remnants of old crater walls often render it possible to reconstruct the former situation."

compound volcano
Schieferdecker (1959), p. 239, term 3991: "A volcanic edifice, having several tops and craters. This may be the result of the shifting of the vent over small distances, thus disturbing the regular form of the mountain, or several young cones may rise above the ruin or [rim] of an older volcano."

conca
Schieferdecker (1959), p. 246, term 4084: "A term introduced by H. Tanakadate (1914) for a conchoidal, lateral subsidence, bounded by gradually sloping walls, and being the result of eruptive activity in the adjoining region." Sakurazima, Japan and Pilomasin, Sumatra, are examples.

cone
[A hill built up of ejecta or salts originating from a volcanic vent or the outlet of a geyser respec-

tively.]

cone basalt
Schieferdecker (1959), p. 237, term 3961: "Basalt differing from plateau basalts by a lower iron content and a [higher viscosity], thus giving rise to the formation of shield volcanoes (Washington, 1922)."

cone-in-cone
Swayne (1956), p. 39: "A new cone built up inside the rim of an old crater."
Preferred synonym: "nested crater."

congealed crust
Schieferdecker (1959), p. 251, term 4162: "The frozen surface of a lava stream or of a lava lake."

consecutive caldera
[Nested caldera.]

corda
[Parallel ridges or corrugations of lava <1 meter in amplitude.]

corded pahoehoe
[Ropy lava, the surfaces of which have numerous parallel to subparallel wrinkles <1 meter in amplitude.]

corded lava
[Lava the surfaces of which have numerous parallel to subparallel wrinkles <1 meter in amplitude.]

core
Webster 3 (1967), p. 506: "The plug or neck of a volcano."

cored bomb
Macdonald, G. in Hess and Poldervaart, (1967), vol. 1, p. 48: "A volcanic bomb formed by the wrapping in liquid lava of solid nuclei. Cores may be fragments of sedimentary, metamorphic, or igneous rocks, or related volcanic rocks, or already solidified lava of the same eruption. The bomb may have many kinds of external forms, some are quite angular."

coulee
Thornbury (1954), p. 496: "An individual lava flow from the crater or flank of a cone forming a tongue-like extension down the cone."

coulee lake
[A ponded portion of a lava flow along its route

from the source vent.]

cow-dung bomb
Schieferdecker (1959), p. 258, term 4257: "A flat, vitreous and porous bomb with a scoriaceous surface, still being [partially] fluid [on impact] [Lacroix, 1930]."

cr   crater
[Any roughly circular depression formed by any means. In volcanology, the term can apply to a caldera, a volcano summit pit, a maar, a lava sink, an ebullition pit or a volcanic bomb impact.]

crateral
Webster 3 (1967), p. 531: "Of or belonging to a crater."

*   crateral magma
Schieferdecker (1959), p. 227, term 3826: "Magma appearing at the surface during an eruption. At depth already phenocrysts have formed and gases have escaped (Perret, 1924)."

crater basin
Stamp (1961), p. 136: Webster: "A depression containing craters."

crater cone
[A symmetrical hill of volcanic origin containing a summit crater.]

crater fill
[Material within a crater. In volcanology, the fill may be pyroclastic fallback or lava or evaporites.]

crater fumarole
Schieferdecker (1959), p. 274, term 4470: "A fumarole [within] a crater (von Wolff, 1914)."

crater island
Schieferdecker (1959), p. 238, term 3981: "A crater, the summit or summit ring of which, rises above the water."
See also "caldera island."

crater lake
[Water filling the floor or portion of the floor of a crater.]

crater-lake lahar
Challinor (1964), p. 140: "Mud flows resulting from the overflow and evacuation of crater-lakes."

crater remmant
Swayne (1956), p. 21: "The remains of a volcanic cone after a paroxysmal eruption which leaves a circular cavity surrounded by low walls." [Originally termed basal wreck.]

crater ring
Schieferdecker (1959), p. 241, term 4026: "The ring–shaped [rim of a crater]."

croicolitic fumarole
Schieferdecker (1959), p. 275, term 4490: "According to L. Palmieri, about 1907, a fumarole arising from a partly cooled-down lava." Cf. "leucolitic fumarole."

crumble breccia
Schieferdecker (1959), p. 261, term 4292: "A breccia composed of lava blocks of different dimensions, crumbled from a lava dome or spine, and accumulated in chaotic fashion. The large openings between the blocks may gradually be filled with finer material."

cryptovolcanism
[Volcanic phenomena possibly at the base of the crust or in the mantle, the effects of which do not necessarily bring volcanic material to the surface but do cause surface structure disruption.]

crystal lapilli
Schieferdecker (1959), p. 256, term 4234: "More or less well-shaped crystals already formed in the magma and explosively ejected in great quantities."

crystal lithic tuff
Schieferdecker (1959), p. 259, term 4273: "Blyth (1940), a tuff in which both crystal and lithic fragments are present [but where] crystal fragments predominate."
Cf. "lithic crystal tuff," and "crystal tuff."

crystal tuff
Stokes and Varnes (1955), p. 35: "An indurated volcanic tuff dominantly composed of crystal fragments and crystals blown out during a volcanic eruption. It may be advisable to restrict the term to tuffs containing more than 75 percent by volume of crystal and to adopt the term crystal-vitric where the content of crystals ranges between 50 and 75 percent (Wentworth and Williams, 1932)."

crystal-vitric tuff
"Tuff consisting of 50–75 percent crystal fragments and 25–50 percent glass fragments."

Cf. "crystal tuff" and "vitric tuff."

cu cumulodome
"A mass of viscous lava protruded from a volcanic vent with little lateral spreading."

cumulovolcano
Webster 3 (1967), p. 553: "A dome-shaped volcano formed by the extrusion of highly viscous lava."

cupola
Halliday (1966), p. 378: "[In volcanology] an indentation or cavity in the ceilings of some lava tube caverns."

curtain of fire
Schieferdecker (1959), p. 236, term 3957: "(G. Macdonald), a wall of serried fire fountains, emitted from an erupting fissure or explosion graben."

cy cylindrical vent
(Perret, 1924): "[A tube created] by the reaming out of a conduit under high pressure and consequent high velocity [of] a flow of gas, charged with [much] ash and lava blocks, torn out from the conduit walls."
Less-preferred synonym: "eruption cylinder."

da dagala
See "steptoe."

de death "valley"
[A depression (crater floor, valley) filled with heavy toxic volcanic gases such as CO_2, CO, H_2S, SO_2.]

debris-cone
Stamp (1961), p. 148: "A cone formed by the accumulation of volcanic ejecta and debris (Dana, 1890)."

defluidization
[The release of fluids—gases, liquids, or melts—to the surface of a cosmic body at any given rate or time.]

degassing phenomenon
See "defluidization."

degassing
[In volcanology the process of release or emis-

sion of gases from volcanic substances.]

dermolith
Johannsen (1939), p. 248: "(Jaggar), pahoehoe."

dermolithic solidification
Schieferdecker (1959), p. 228, term 3833: "Freezing of the surface of pyromagma."

deroofing
"Fusing or disruption of the roof of a batholith leading to its assimilation or engulfment resulting in an extrusion or areal eruption."

*   desmosite
Holmes (1928), p. 77: "Jaggar (1917), a term, meaning 'skin-stone' applied to ropy-lava or pahoehoe-lava."

detonation
[An explosion made by the combustion of volatile gases, or by the abrupt release of gases from a vent.]

detritus eruption
[An eruption producing much pyroclastic debris.]

di diatreme
Stokes and Varnes (1955), pp. 38–39: "A volcanic vent produced mainly by gaseous explosions. The tube or vent may be filled after the explosive phase with a heterogeneous breccia of materials from below, from the sides, or even from the surface. No specific size is implied and some diatremes are only a few feet across."

dike phase
"The closing episode in a volcanic cycle, characterized by the injection or minor intrusions, especially dikes."

dilation
[Process of widening or enlarging.]

discoid
Cook (1966), ESR, vol. 1, p. 164: A term used to describe the welded portion of a ignimbrite up to a meter, and as much as 5 cm in thickness.

do dome
Holmes (1928), p. 84: "A rounded extrusion of

highly viscous lava squeezed out from a volcanic vent, and congealed above and around the orifice instead of flowing away in streams. Portions of older lavas or ejectamenta may be elevated by the pressure of the new lava rising from beneath."

dome volcano
Stamp (1961), p. 161: "A volcano composed of very viscid lava, which, on eruption, accumulates dome-shaped above the vent, leaving no crater."

domical protrusion
[A dome.]

dormant
Webster 3 (1967) p. 1275: "That which is quite inactive, as though sleeping, but which may be awakened later into significant activity or effect."

double crater
Schieferdecker (1959), p. 240, term 4015: "Twin crater."

dr   Schieferdecker (1959), p. 252, term 4173: "Pertaining to a cover [or veneer] of dripping lava along the walls of a lava tunnel."

driblet
[Spatter that has flowed after landing, forming pendants, sagged droplets, stalactites, etc.]

driblet cone
Stokes and Varnes (1955), p. 43: "A miniature lava cone formed by the accretion of drops of lava projected from steam vents or blowholes and falling [around the vent, or to one side in the case of asymmetric venting]."
Synonym: "spatter cone."

driblet spire
Macdonald, G. in Hess and Poldervaart, (1967), vol. 1, p. 55: "A relatively tall thin column of agglutinate built on the surfaces of lava flows by the escape of gas and clots of molten lava through cracks or other openings in the crust of the flow. They grade downward in size to hornitos. Some straight columns are less than a meter in diameter and as much as 4 m high, although most are less than 2 m high."

drum fire
Schieferdecker (1959), p. 265, term 4361: "[In volcanology] a prolonged series of continuous

explosions."

dry maar
Schieferdecker (1959), p. 244, term 4058: "A wide crater, more or less at the level of the landscape, with a dry and rather flat bottom and surrounded by a low ring wall only. Several of these dry maars are to be found around Mt. Lamongan in East Java (Traverne, 1926)."

du   dust cloud
Ross and Smith (1961), p. 4: "Airborne pyroclastic material of dust size that characterizes explosive volcanic eruptions. The dust clouds associated with ash flows are not basically different from glowing clouds, glutwolken and nuées ardentes although these terms emphasize the glow that is normally reflected from the underlying incandescent ash flow."

dust-tuff
Fisher (1966), p. 294, modified after Bailey (1926): "Pyroclastic rock showing no effect of erosion in which component fragments are less than 50 percent of the rock. The diameter of fragments is less than [0.05] mm."
Synonym: "mud-tuff."
Cf. "tuffaceous shale."

dy   dyngja (pl. dyngjur)
Stamp (1961), p. 168: "A gently-sloped volcano formed by successive outpourings of fluid lava unaccompanied by accumulations due to violent ejection (Iceland)."

eb   ebullitim
Webster 3 (1967), p. 718: "The act, process, or state of boiling or bubbling up. A sudden and violent outburst or display."

ebullitim crater
[A small crater formed on the surface of a flow by gases rising in it. Observed on the 1951 Mt. Lamington nuée ardente.]

ec   eccentric eruption
Schieferdecker (1959), p. 266, term 4372: "An eruption emanating from an adventive crater."

ef   effluent lava flow
"A lava flow that is discharged from a volcano by way of a lateral fissure."

effusion
Webster 3 (1937), p. 726: "The action or process of effusing or of being poured out."

effusiometer
Webster 3 (1967), p. 726: "An apparatus for determining the effusion velocities of gases and hence their densities."

effusive
Challinor (1964), p. 80: "Pouring, or poured, out. Applicable to lavas and nuées ardentes."

effusive rocks
Johannsen (1939), p. 174; "[Rocks solidifying from melts or emulsions] that have been poured out upon the surface of the earth."
Synonyms: "extrusive rocks," "volcanic rocks."

ej   ejecta
Webster 3 (1967), p. 729: "Material thrown out [as from a volcano]."

ejected blocks
U.S. Bureau of Mines: "The larger fragments of a volcanic breccia, generally derived from the internal or subjacent rocks of a volcano, and often highly metamorphosed."
† [Material larger in diameter than 32 mm thrown out of a volcanic vent.]

ejectamenta
Holmes (1928), p. 87: "A general term for pyroclastic materials ejected from a volcanic vent; classified by Johnstone-Lavis as 'essential' when they consist of material directly derived from the magma of the eruption; 'accessory' when they consist of re-ejected portions of the volcanic cone; and 'accidental' when they consist of older rocks underlying the volcano (Johnston-Lavis (1886) p. 421)."

ejection
Schieferdecker (1959), p. 236, term 3949: "Throwout of lava as incoherent pieces; according to Reck (1915)."

el   elephant-hide pahoehoe
[Ropy lava with a surface texture resembling the skin of an elephant.]

ellipsoidal
Holmes (1928), p. 88: "A structural term applied to spilitic and similar rocks as a result of the conditions under which they consolidated.

They are disposed in a series of sack-or pillow-like masses."
See "pillow structure."

em   embryonic dome
[A lava cone of moderate height over vents producing enormous outflow of basalt.]

embryonic eruption
[Eruptive energy producing volcanic tremors but no surface venting.]

embryonic volcano
[An incipient, newly created volcano in the process of development.]

emulsion rock
Walker and Skelhorn (1966), p. 97: "An intimate mixture of acid and basic glasses" as at Breiddalur, eastern Iceland; p. 99: "In these occurrences there is evidence that two magmas were erupted simultaneously from the same vent."

en   endogenous dome
Wentworth and Williams (1932): "A volcanic dome that has grown primarily by expansion from within and is characterized by a concentric arrangement of flow layers."

end moraine
Schieferdecker (1959), p. 252, term 4176: "Of scoriae, the frontal moraine."
See also "scoria moraine."

engulfment
Monkhouse (1965), p. 110: "The inward collapse of a volcanic [structure], the result of molten lava beneath it being drawn off under the surface of the earth or through a fissure in the flanks of [the structure]. This is one way in which a caldera may be formed." [Collapse may also be achieved by phase change, by solidification of a magma chamber, etc.]

entrail pahoehoe
Macdonald, G., in Hess and Poldervaart, vol. 1, p. 12: "Consists of a heap of long narrow contorted pahoehoe toes jumbled together and giving the impression of being intertwined, resembling heaps of intestines from slaughtered animals;" p. 13: "formed by the extrusion of successive dribbles of lava through cracks in the underlying flow crust;" p. 12: "each of the dribbles forms a small toe, which drapes itself

over and around the earlier toes."

ep   epiclastic tuff
*   Fisher (1966), *ESR*, vol. 1, p. 289: "Term is a 'contradiction' because 'tuff is composed of pyroclastic fragments.' The fragments are not broken by weathering or erosion as the word 'epiclastic' implies."

epigenetic volcano
Schieferdecker (1959), p. 239, term 4001: "A volcano formed at a later date; according to Reck (1930), a small body formed by an ephemeral activity in the environs of the already extinct crater or volcano as a last impulse of a nearly exhausted magma mass."

epigenetic volcanism
Schieferdecker (1959), p. 239, term 4000: "Of a later date than that of the main period of activity; accordingly to Reck (1930), an after-effect of a nearly exhausted magma mass."

er   erosion caldera
[A caldera modified by erosion.]

*   eructation
Webster 3 (1967), p. 773: "A violent belching out or emitting as of gaseous or other matter from the crater of a volcano."

erupt
Webster 3 (1967), p. 773: "To force out or release [as something pent up], usually suddenly and violently; cause to erupt; throw out; expell, eject as 'the volcano erupted lava bombs'."

eruption
Monkhouse (1965), p. 114: "The process by which solid, liquid or gaseous materials are extruded or emitted onto the earth's surface as a result of volcanic activity . . . The materials erupted include gaseous compounds of sulphur, hydrogen and carbon dioxide; steam and water-vapour; scoria, pumice, cinders, dust, ash; acid and basic lava [hence 'eruptive rocks']. The eruptions may range from quiet outflows and outwellings, to explosive and paroxysmal activity."

eruption channel
Schieferdecker (1959), p. 243, term 4046: "Volcanic vent."

eruption cloud
Schieferdecker (1959), p. 268, term 4405: "A column of gases, ash and rock fragments, [emitted] from a crater or nuée ardente, forming a cloud which may rise to great heights. Those of Krakatau rose to a height of more than 50 km in August 1883. Eruption clouds can have shapes resembling pine trees (term 4409) and cauliflowers (4410)."
Less-preferred synonym: "explosion cloud."

eruption column
Schieferdecker (1959), p. 268, term 4407: "The lower, column-shaped, vertical part of the eruption cloud, where the ash and debris-laden gases are still rising with tremendous speed."

eruption cone
Schieferdecker (1959), p. 237, term 3971: "Cone formed by eruptions."

eruption lahar
[A hot, volcanic mud flow.]

eruption cycle
Schieferdecker (1959), p. 267, term 4381: "The regular change in the behavior of eruptions in the period of activity of the volcanic cycle. Perret (1924) and Neumann van Padang (1931) distinguished various phases."
See also "volcanic cycle."

eruption cylinder
Schieferdecker (1959), p. 243, term 4049: "A cylindrical vent."

eruptivity
Webster 3 (1967), p. 773: "The state of being eruptive—return of the geyser from a dormant phase to eruptivity."

eruptive focus
Schieferdecker (1959), p. 263, term 4322: "The explosion focus."

eruptive fumarole
"Gases escaping explosively from an opening in a lava flow or from fractures in the volcano flank, dragging with them clots of lava, which may form a driblet cone."

eruption rain
Schieferdecker (1959), p. 269, term 4420: "Rain [induced] by a [volcanic] eruption, and which would not have fallen otherwise."
Less-preferred synonym: "volcanic rain."

es   essential
[Pertaining to volcanic material identical in

composition to the bulk of the associated flow or pyroclastics.]
Synonym: "indigenous."

essential ejecta
Macdonald, G., in Hess and Poldervaart, (1967), vol. 1, p. 51: "Fragments of a crust or other solidified lava formed in the vent during the same eruption and later disrupted."

essential tuff
Nelson and Nelson (1967), p. 128: "A pyroclastic rock composed of debris [and identical in composition with the] erupting lava.
Cf. "accessory tuff," "accidental tuff."

eu   eutaxite
Ross and Smith (1961), p. 4: "Fritsch and Reiss (1868, p. 414) proposed the name eutaxite for a volcanic rock composed of ejected fragments of different colors, and texture as follows: 'The different fractions in general lie beside one another as streaks, bands, and lenses in seemingly well ordered distribution.' Piperno seems to be a similar rock but may however, differ slightly by having larger lenses of glass known as fiamme."

ex   exhalation
[Fluid given off.]

exogenous dome
Wentworth and Williams (1932): "A volcanic dome that is built by surface effusion of viscous lava, usually from a central vent or crater."

explosion caldera
Stokes and Varnes (1955), p. 49: "A caldera created by explosive eruptions that have removed large masses or rock."
Cf. "subsidence caldera."

explosion cloud
Schieferdecker (1959), p. 268, term 4445: "An eruption cloud."

explosion crater
H. Stearns, (personal communication): "Crater formed chiefly, or entirely, by explosion."
Synonym: "explosion vent."

explosion focus
Schieferdecker (1959), p. 263, term 4322: "A point in the top of the magma column, directly underneath the solidified plug which acts as

a choke for the accumulating volcanic gases (van Bemmelen, 1949)."
Less-preferred synonym: "eruptive focus," "volcanic focus."

explosion funnel
Schieferdecker (1959), p. 241, term 4018: "The conical depression in the top of a volcano."

explosion graben
Schieferdecker (1959), p. 236, term 3956: "Long, narrow sink with deep walls, originating when a great many closely spaced explosion vents are arranged in a long row (von Wolff, 1914)."

explosion lake
[A water-filled maar.]

explosion level
[The horizon in the upper part of a magma column where gases nucleate. The level will fluctuate as a function of the confining pressure and gravity.]

explosion pit
Longwell, et al., (1969), p. 649: "A vent, drilled to the surface by volcanic gases, from which no lava issues."

explosion point
Schieferdecker (1959), p. 265, term 4356: "The physicochemical stage at which the gas content of the magma escapes explosively." [Retrograde boiling, pressure drop, temperature increase, or addition of gas from below can trigger the explosion point.]

explosion vent
[A vent marked by volcanic explosivity.]

explosion tuff
[A tuff produced by explosivity at the source vent.]

exploding bomb
See "explosive bomb."

explosive bomb
Schieferdecker (1959), p. 258, term 4253: "A kind of breadcrust bomb which, during the flight or after falling, violently throws off fragments of its thick glassy crust in consequence of tensions set up by cooling and retrograde boiling of its interior (A. Rittmann)."
Less-preferred synonym: "exploding bomb."

explosive eruption
Stokes and Varnes (1955), p. 49: "An eruption characterized by violent explosions and the production of ashes and cinders."

explosive index
See "explosivity index."

explosivity index (E)
According to Rittmann (1944):
Explosivity Index E=100×mass of fragmental material ejected ÷ total ejected material.
Rittmann in Runcorn (1967), vol. 2, p. 824: "E less than 10 is low explosivity, and is 'linked to oceanic and continental basaltic magmas'— E greater than 90 is high explosivity and 'characterizes acid magmas of orogenic volcanism'," Synonym: "explosive index."

* external volcanism
Schieferdecker (1959), p. 221, term 2729: "All [volcanic surface] phenomena such as effusions of lava, explosions and the ejections of ashes and rock or the emissions of volcanic gases." Synonym: "extrusive volcanism" [preferred].

extinct volcano
Schieferdecker (1959), p. 224, term 3782: "No longer showing any signs of volcanic activity and of which no eruptions are expected, its intercrustal magma being congealed or nearly so and its supply of gas being terminated."

extravasate
Webster 3 (1967), p. 807: "To cause to pour out or erupt as of lava from a vent in the earth."

extravasation
Webster 3 (1967), p. 807: "The action of extravasating; the condition of being extravasated as the extravasation of lava from a volcano; effusion, eruption; an extravasated fluid or a deposit formed from extravasation; extravasate."

extrude
Webster 3 (1967), p. 808: "To move to or appear at the surface or the outside: emerge as lava extruding from early fissures."

extrusion
"In volcanology, the emission of magmatic material [generally lava] at the earth's surface; also the body of rock or the structure produced by the process, such as a lava flow, a layer of pyroclastic rock, or a volcanic dome."

extrusive
Stokes and Varnes (1955), p. 50: "Pertaining to igneous material poured out on the surface of the earth in a molten state and to fragmental material of all sizes erupted from volcanic vents. Lava flows and tuff beds are common examples."

ex   extrusive body
Schieferdecker (1959), p. 236, term 3351: "A mass of volcanic material formed by extrusions, as a volcanic cone, a lava flow, etc."

extrusive rock
Geikie (1879): "Extrusive rocks are classified in two great groups: (1) lavas ['those which have been poured out in a molten condition at the surface']: and (2) fragmental materials ['including all kinds of pyroclastic detritus discharged from volcanic vents']." Synonym: "effusive rock" (see "intrusive rock").

fa   fango
Webster 3 (1967), p. 822: "Mud, mire; esp. a clay mud from hot springs at Battaglio, Italy, that is used in the form of hot external applications in certain medical treatments."

farinaceous lava
Schieferdecker (1959), p. 249, term 4128: "Particles of a white hot lava stream as at in Halemaumau which seem to be in a state of mutual repulsion and flowing as meal."

fault vent
Webster 3 (1967), p. 829: "A volcanic vent situated on a fault."

favilla (pl. favillae)
Webster 3 (1967), p. 830: "A small incandescent fragment of lava from a volcano."

fe   feeder
[In volcanology, a cylindrical to planar conduit through which magma is brought to the surface.]

* feeding tube
Macdonald, G., in Hess and Poldervaart, (1967):
Less-preferred synonym: "lava tunnel."

feeding vent
Schieferdecker, (1959), p. 243, term 4046: "A volcanic vent."

fi   **fiamme**
"Fiamme are flattened, black, glassy inclusions with flamelike cross sections originally applied to such structures in piperno but, now generally applied to flattened glass lenticles in welded tuffs. These are usually considered to have formed by the collapse of included molten pumice under the load of the overlying portions of the pyroclastic flow." According to Ross and Smith (1961): "fiamme in ignimbrites are un-expanded lumps of volatile-rich 'very fluid' glass carried up in the vent during an initial eruption. Rittmann (1960, pp. 94–97) believes they are chilled clots of lava. The lenticles are often several centimeters in length but may range from millimeter to meters."

fiery cloud
See "nuée ardente."

filiform lapilli
Schieferdecker (1959), p. 256, term 4235: "Pele's hair."

filamented pahoehoe
[Ropy lava, portions of which are drawn out into fine glass filaments (i.e. Pele's hair).]

fire avalanche
[Glowing avalanche or nuée ardente.]

fire fountain
H. T. Stearns: "A fountain of liquid lava; also called 'lava fountain' or spouts of fire. Rarely used as a verb—fire fountaining."

fire pit
Webster 3 (1967), p. 855: "A pit whose floor is wholly or partly filled with incandescent lava."

fissure
Schieferdecker (1959), p. 246, term 4094: "A fracture in a volcano or in a lava dome, but also traceable in [volcanic terrain] as a deep and nearly straight cleft, usually no more than a few meters in width, but sometimes extending for many kilometers (Hobbs, 1919, p. 247). In Iceland there is a series of [large parallel] fissures, called 'gjas'."

fissure basalt
[Basalt issuing from linear vents or fractures as distinct from basalt issuing from point vents, volcanoes or cones.]

fissure eruption or effusion
Monkhouse (1965), p. 124: "A linear volcanic eruption, in which lava, generally mafic and of low viscosity, wells up to the surface along a line of crustal weakness, usually without any explosive activity. The results may be an extensive basalt plateau." [The Laki fissure in Iceland is 36 km long.]
Cf. "central eruption."
Synonym: "linear eruption."

fissure flow
"Lava which passively flow out of a fissure."
See "fissure eruption."

fissure-flow volcano
"An outpouring of highly fluid lava from a fissure or crack in the earth crust, localizing a volcano [or a series of volcanoes] along it."
Synonym: "fissure volcano."

fissure fumarole
[A fracture the entire length of which serves as a gaseous vent.]

fissure vent
[A vent localized along a volcanic fissure.]

fissure volcano
See "fissure-flow volcano."

fl   **flame structure**
See "fiamme."

flaming orifice
Schieferdecker (1959), p. 275, term 4478: "A fumarole with burning gases, bluish for H or H_2S flames and yellow when containing Na (Williams, 1954)."

flank eruption
Schieferdecker (1959), p. 266, term 4370:" "An outburst on the flank of a volcano, [cone or caldera]."

flank outflow
Schieferdecker (1959), p. 266, term 4371: "Lava flowing out of the flank of a volcanic structure. Dana (1888) termed this 'effluent lava outflow'."
Less-preferred synonym: "effluent lava flow," "transcrateric lava effusion."

flood
Webster 3 (1967), p. 873: "A great stream of something [as light or lava] that flows in a steady course."

flood basalt
Challinor (1964), 3rd ed., p. 97: "[Chemically homogeneous, very fluid] basaltic lava which issues in large quantities, from a fissure or a central arifice, flooding the neighboring country."

flood basalt
Cook (1966), *ESR*, vol. 1, p. 167: "In Columbia Plateau of Washington, Oregon and Idaho, (p. 169) a single flow may be several thousand square miles in extent, although only a few hundred feet thick". The now known great extent of individual basalt flows implies very low viscosity and rapid spreading, or long-continued mobility, perhaps in conjunction with simultaneous feeding from multiple fissures."

flotation bomb
Schieferdecker (1959), p. 258, term 5255: "A lava ball."

flow
Stokes and Varnes (1955), p. 55: "That which flows or results from flowage; as, for example, a 'lava flow'."

flow lava
Schieferdecker (1959), p. 248, term 4110: "Lava of a lava flow."

flow marking
[Patterns on flow surfaces indicating flowage.]

flow rocks
See "flows."

flow structure
[A petrographic texture in which minerals, rock components, inclusions, or cavities show a parallel arrangement or deformed pattern in the flow direction which was produced during their emplacement or formation.]

flow unit
Macdonald, G., in Hess and Poldervaart, (1967), vol. 1: from Nichols (1936): "A successive but essentially contemporaneous layer or portion constituting a single larger flow. Each flow unit represents a separate gush or sheet of liquid lava pouring over one another during the course of a single eruption."

fluent lava
Schieferdecker (1959), p. 249, term 4125: "Lava with a smooth coherent surface, varnished and vitrified to a depth of a few milimeters [as on ropy lava] (Lacroix, 1936)."

Less-preferred synonym: "skin lava," "dermo-lithic lava," "dermolith."

fluidization
Challinor (1964), p. 98: quotes Reynolds (1954): "An industrial process in which gas is passed through a bed of fine-grained solid particles in order to facilitate mixing and chemical reaction. At a particular rate of gas flow the bed expands and the individual particles become free to move. With increase in the rate of gas flow, a bubble phase forms and travels upwards through the expanded bed in which the particles are violently agitated; the bed is now said to be fluidized. With continued increase in the rate of gas flow, more and more of the gas travels as bubbles containing suspended solids, until ultimately the solid particles become entirely entrained and transported by the gas."
Nelson and Nelson (1967), p. 182: "A bed of coal is fluidized when it is made to float by the upward movement of a current of liquid or gas. In such a bed, friction between particles is zero and they become highly mobile. Fluidization is used in the calcination of various minerals, the coking of petroleum pitch, in Fischer-Tropsch synthesis, and in the coal industry."
Holmes (1965), p. 271: "Tuffisite pipes show effects of fluidization. Rounded particles brought from great depth indicate that 'swiftly moving gas streams charged with particles provide an efficient means of transport'. Gas acting as an erosive and transporting agent, i.e., fluidization erosion, is probably the chief mechanism responsible for the chimney-like form of volcanic pipes. Fluidization involves erosion, transport and chemical action by streams of gas."

fluidize
Webster 3 (1967), p. 877: "To cause to behave like a fluid; specif: to suspend [a finely divided solid] in a rapidly moving stream of gas or vapor so as to induce flowing movement of the whole [as in the transport of flour by air blast or in the catalytic cracking of petroleum]."

fluidized
Longwell, et al., (1969), p. 650: "An adjective describing a body of solid particles in a dilatant state; that is, the particles are no longer in continuous contact with one another."

fluviovolcanic
[Pertaining to water-sedimented volcanic material.]

fo   foamy lava
Schieferdecker (1959), p. 249, term 4127:
"Very brittle lava in the crater Dolomieu of
Piton de la Fournaise; lava with regular vesicles, the separating walls of which are very thin
and more fragile than those of pumice. It
breaks by pressure of the fingers. The crust of
the lava is rough, showing holes and peaks of
glass (Lacroix, 1936)."
Synonym: "reticulite," "pumiceous lava,"
"thread-lace lava."

fosse
Schieferdecker (1959), p. 241, term 4023: "An
atrio or moat surrounding a volcanic structure
within a larger volcanic structure."

fountain
Webster 3 (1967), p. 898: "In volcanology, an
upward, sideward or downward jet of ash, lava
or mud."

fr   fragmental
Nelson and Nelson (1967) p. 149: "A term
sometimes used to describe the texture of volcanic tuffs and breccias, which represent the
solidified deposits of pyroclastic materials
erupted by volcanoes."
See "pyroclastic."

frozen lava
Schieferdecker (1959), p. 226, term 3808:
"Lava in a solid state."

fu   fumarole
Swayne (1956), p. 63: "A vent in a [volcanic
province, violently or passively] giving off a
chemically-active fluids which frequently alter
the character of the surrounding rocks. The
nature of these fluids varies with temperature
and allows a classification into. . . ."
1. dry fumarole evolving anhydrous chlorides,
   above 630°F.
2. acid fumarole evolving hydrochloric acid
   and sulfur dioxide at temperatures above
   180°F.
3. alkaline fumarole evolving steam and am-
   monium chloride at temperatures above
   180°F.
4. cold fumarole evolving nearly pure steam,
   above 180°F.
5. mofette evolving carbon dioxide, nitrogen,
   oxygen at atmospheric temperatures.
See also: "crater fumarole," "primary fumarole," "secondary fumarole," "tertiary fumarole," "eruptive fumarole," "roaring fuma-

role," "chlorine fumarole," "acid fumarole,"
"alkalic fumarole," "cold fumarole," "fissure
fumarole," "tunnel fumarole," "leucolitic
fumarole," "croicolitic fumarole," "steam
fumarole."

fumarole field
[An areal grouping of fumaroles arranged on
zones of weakness or sometimes in a random
pattern.]

fumarole mound
Schieferdecker (1959), p. 276, term 4505: "A
small mound in which the fumaroles or solfataras occur as at Kawah or as in the crater of
Papandajan (Java) . . . Allen and Zies (1923)
attributed such hillocks to [differential] de-
flation by the wind or to accumulation caused
by small eruptions."
Cf. "solfatara mound."
Synonym: "fumarole pimple."

*   **fumarole pimple**
Schieferdecker (1959), p. 276, term 4505:
"fumarole mound."

fumarolic
Webster 3 (1967), p. 920: adj. form of "fumarole."

fumarole stage
[The period of activity of a volcanic structure
characterized by significant or marked gaseous
emission from vents on the floor, rim, or flank.]

fume
Schieferdecker (1959), p. 274, term 4463:
"[Gaseous emission] which may contain dust,
rising from the crater, from fumaroles or solfataras, or from lava which flows into water
(sea or lake)."

furrowed volcano
Schieferdecker (1959), p. 247, term 4104:
"Sapper's (1927) term for the first stage of destruction of a volcano by erosion, the flanks
being cut by valleys which do not yet disturb
the regular shape of the cone."
Cf. "cleaved" volcano and "volcanic wreck."

fusiform bomb
[A volcanic bomb which was in a molten state
when ejected (as distinct from solid blocks).]
"Fusiform bombs commonly range in size up
to 1–2 m in length and more than 1000 kg
weight." (Wentworth and Macdonald, 1953,
p. 81.)

ga **gas fluxing**
Schieferdecker (1959), p. 226, term 3812: "[A process which] presumes a blowpiping activity of the gases escaping from the pyromagma. Their heat [without or with the help of additional heat from gas reactions] may be sufficient to melt superficial frozen lava as well as the walls of craters. Fissures may thus be enlarged so as to form pipes (Daly, 1914)."

gas emission phase
Schieferdecker (1959), p. 268, term 4394: "A paroxysmal emission of gas, the predominant feature being the ejection of incandescent liquid lava with a preponderant proportion of ash. The liberation of highly compressed vapor and gas most strongly characterizes this phase (Perret, 1924)."

gas seepage
Schieferdecker (1959), p. 277, term 4519: "The emanation of gas from the surface more or less diffusely, i.e., in a dispersed manner, not one or more distinct openings."

gi **giant's causeway**
Schieferdecker (1959), p. 253, term 4185: "A more or less horizontal surface made of prismatic columns due to contraction by cooling, resembling a kind of platform."

gj **gja**
"A fissure from which volcanic eruptions take place. They may be filled with pyroclastics and lava as in the Norway-Greenland basin" (Gakkel and Dibner in Runcorn, 1967 ,vol. 1, p. 160.) Icelandic, for "chasm."

gl **glaze**
Halliday (1966), p. 379: "A shiny, relatively smooth coating of some lava tube walls apparently the result of the action of hot gases." [The term can also be applied to walls of any volcanic surface so coated or fused.]

glowing avalanche
Schieferdecker (1959), p. 272, term 4451: "An avalanche of the gas-rich suspension of either: (a) great masses of incandescent blocks, lapilli ash and dust, (b) the debris of a disintegrating lava spine, (c) the [lower fragmental debris layer] of a glowing cloud (nuée ardente) that follows the depressions in the landscape."
Synonym: "fire avalanche," "hot avalanche,"

"volcanic avalanche" (Ross and Smith 1961, p. 4).

glowing-avalanche deposit
"Chaotically brecciated material deposited by a 'glowing avalanche" not to be confused with 'welded tuff'."

gl **glowing cloud**
"A gas-generating, eruptive phenomenon of which the heavy fractions of incandescent debris avalanche and follow the depressions on the flank of the caldera or volcano, and spread over the adjacent landscape, while the lighter fractions of volcanic gases, ash and dust cauliflower upward." [Thus, a nuée ardente includes a glowing avalanche but a glowing avalanche does not necessarily include a nuée ardente.]
Synonym: "nuée ardente" (preferred); "Glutwolke."

gr **grooved lava**
Macdonald, G., in Hess and Poldervaart, (1967), vol. 1, p. 35: "Semisolid but still-plastic lava, either aa or pahoehoe, moving past an irregular surface of more rigid lava or other rock becomes grooved and striated. When the lava freezes, the grooved surfaces commonly are preserved and resemble coarse-textured slickensides. However, they differ from slickensides or glacially striated surfaces in being rough or granular, or even somewhat spinose. Commonly the tiny projections on the surface are asymmetrical . . ."

grooved squeeze-up
[A squeeze-up (q.v.) which in the process of formation displays nearly vertical grooves on the "fin" surfaces.]

ground moraine
[In volcanology, a basal moraine of lava debris.]

gu **gust**
Schieferdecker (1959), p. 272, term 4449: In volcanology, a violent blast of wind, preceding a nuée ardente.

ha **harrah**
Stamp (1961), p. 145: "Tracts of rough lava surfaces in Arabia which cut the feet of man and animals."

Hawaiian type eruption
Schieferdecker (1959), term 4334: "Hawaiian type volcanic activity is characterized by very fluid, effusive basaltic lava, sometimes forming an incandescent lake with low gas pressure. It may give rise to lava flows and fountains but explosive phenomena are rare."

he   **helluhraun**
Schieferdecker (1959), p. 249, term 4122: "Icelandic term for 'pahoehoe'."

hemidiatreme
"A magma column which did not reach the surface but domed the overlying beds."
Synonym: "A laccolithic dome" [preferred].

ho   **homates**
*   Entry in a book reivew (by A. Johannsen) of Schneider (1911): "A type of volcano belonging to 'klasmatitic' groups."
Synonym: "rheumatitic," "rheuklastitic," "klasmatitic."

hornito
Macdonald, G., in Hess and Poldervaart, (1967), vol. 1, p. 55: "A 'little oven'. It is a small cone of agglutinate built on the surfaces of lava flows by the escape of gas and clots of molten lava through cracks or other openings in the crust of the flow. Some are broadly rounded and in shape resemble the domical baking ovens used by the Indians in Central America. The term was first used by Alexander von Humboldt (1823); cited by K. Sapper (1919), p. 1. Most have a hollow center. They range in size up to several meters in height and in basal diameter." [and can grade into lava spires.] [The term should be reserved for structures built at "rootless" vents on lava flows.]

horseshoe-shaped crater
Schieferdecker (1959), p. 241, term 4030: "A breached crater."

hot avalanche
Schieferdecker (1959), p. 272, term 4451: "A glowing avalanche (q.v.)."

hot mudflow
Schieferdecker (1959), p. 271, term 4435: "Volcanic mudflows (1) "result from eruptions breaking through a crater lake, as during the Kelud and St. Vincent eruptions, when the crater water was ejected and mixed with hot material of the eruption [eruption lahar]," (2) "result from heavy rain falls following strong ash and block eruptions, avalanching the hot material downward [rain lahar]."
Synonym: "hot lahar."

hot lahar
Schieferdecker (1959), p. 271, term 4435: "A hot [volcanic] mudflow (q.v.)."

hot pool
Schieferdecker (1959), p. 277, term 4520: "A pool or pond filled with water [emitted] from a hot spring." [Bubbling gases or extremely high subsurface temperatures can produce a boiling pool, spring or lake.]
See "geyser pool."

hr   **hraun**
"Icelandic term for recent lava flow."

hv   **hverir**
Schieferdecker (1959), p. 278, 4536: "Icelandic boiling springs including geysers (Thorarinsson, 1955)."

hy   **hyaloclastite**
Cited by Macdonald, G., in Hess and Poldervaat, (1967), vol. 1, p. 30: Term proposed by Rittmann (1960) p. 82: "Hydrated tuff-like rocks formed by 'granulation of lava flows in water' typically composed wholly of angular fragments ranging from 1–2 mm to a few cm across composed largely of yellowish-brown earthy or waxy palagonite. Fragments are generally flattish chips and flakes generally cemented by interstitial zeolites and/or calcite commonly completely massive with no sign of stratification or sorting—p. 31: formed by granulation due to quenching of the front of a lava flow advancing into or beneath water."

hyaloclastic
Rittmann (1962) p. 72–73: "Pertaining to a hyaloclastite (q.v.)."

hydroexplosion
H. T. Stearns "Any explosion caused by contact of lava with water."

hypogene spring
Schieferdecker (1959), p. 287, term 4534: "A spring, sometimes pulsating, in which the water comes from a great depth, and has a distinct

relation to volcanic phenomena (Suess, 1902; Haug, 1927)."

ig   igneoaqueous
*   [Ambiguous term.]

ignisept
Schieferdecker (1959), p. 228, term 3821: "A dike-like magma chamber, but on a much larger scale, extending for hundreds of miles and underlying whole mountain ranges or volcanic chains (Jaggar, 1947)."

ignimbrite
† "An extrusive rock unit make up of glass shards, crystals and lithic fragments commonly in that order of decreasing abundance."
Cook (1966), p. 158: In New Zealand the sheet rhyolites of the North Island were named ignimbrites [fiery raincloud rocks] by Marshall (1932) p. 198–202: "Most students of ignimbrites believe them to have been formed by deposition from a hot, rapidly expanding, turbulent highly mobile, magmatic gas [glass-dust or ash emulsion; nuée ardente or pyroclastic flow] which carries with it intra-telluric crystals, liquid droplets of the exploding magma [and the resulting glass shards], as well as rock fragments torn from the walls of the vent or picked up from the ground surface." [The deposit is generally but not exclusively silicic and often but not necessarily welded.] Other zones within ignimbrites [vapor phase etc.] have been recognized by Smith (1960) and others.
Synonym: "Ash flow sheet" (Smith 1960), "Ash flow cooling unit" (Smith 1960).

ignispumite
Cook (1966) p. 161: "Certain lenticulate to banded rhyolitic rocks of Hungary have been called ignispumites by Panto (1962 p. 307–331) who believes they formed from a heterogeneous acid 'foam-lava'; these ignispumites are highly variable in appearance, contain 'stripes and shards fluidally arranged in two or more vitric phases', and show transitions into true ignimbrites."

im   imbricated volcano
Schieferdecker (1959), p. 237, term 3968: "A volcano built up of sheets overlapping each other like tiles of a roof, younger layers being intruded between older ones (Russel, 1897)."

in   inactive volcano
Longwell, et al., (1969), p. 662: "A volcano that has not erupted within historic time."

inbreak crater
Schieferdecker (1959), p. 241, term 4019: "A pit crater."

incandescent
Webster 3 (1967), p. 1140: "White, glowing, or luminous with intense heat as an incandescent ash flow."

incandescent-ash flow
Schieferdecker (1959), p. 273, term 4457: "Fiery gas-saturated flow, consisting mainly of fine material, in which lumps of pumice and numerous small bits of sedimentary and volcanic rock are mingled rather promiscuously (Fenner, 1923, 1937) originating from a crater or from fissures." [An incandescent or fiery avalanche can be an incandescent ash flow but an incandescent ash flow is not necessarily an incandescent avalanche.]
Synonyms: "incandescent pumice flow," (no q.v.), "incandescent tuff flow" (q.v.), "hot tuff flow" (no q.v.), "hot sand flow" (no q.v.).

incandescent detritus
Schieferdecker (1959), p. 255, term 4210: "Volcanic ejecta thrown out in a glowing state (Perret, 1924)."
Cf. "nonluminous detritus."

incandescent-tuff flow
Ross and Smith (1961), p. 5, Fenner (1948), p. 879: "Relative to the 'tuff flows' of the Arequipa region of Peru, the deposits are the result of a series of fragmental outbursts of rhyolitic lava, similar to that which occurred in the Valley of Ten Thousand Smokes in Alaska."; also stated (1948), p. 882, "an especially important feature . . . is the ability of such tuff flows to spread widely over level or gently inclined surfaces." Jenks and Goldick (1956), p. 156 use the term 'tuff flows' for the same deposits."
Synonym: "incandescent ash flow."

indirect eruption
[An explosion resulting from sudden release of entrained magmatic gases by heating of congealed gas-charged glassy rocks in a volcanic vent.] These gases according to von Wolff (1914) may have led to the "indirect" eruption of Bandaisan in Japan on July 15, 1888.

in   intercolline
Stamp (1961), p. 262: "In volcanology pertaining to a hollow lying between conical hillocks made up of accumulations from volcanic eruptions."

interfluent lava flow
"A lava flow that is discharged into and through subterranean fissures and cavities in a volcano. They may never reach the surface (Dana, 1890)."

internal melting hypothesis
Schieferdecker (1959), p. 226, term 3813: "In volcanology, the concept that the summit of a volcanic cone might collapse as a result of large scale internal melting, taking place at depth as von Hochstetter and Griggs suggested, or within an open crater as Fenner supposed."

intertrappean
Webster 3 (1967), p. 1183: "Lying between successive basaltic lava flows."

intrusive tuff
"Tuffisite"

intumesence
Schieferdecker (1959), p. 242, term 4035: "A vaulted [or inflated] part in the [molten or congealed] crust of a lava flow or of lava on the bottom of a crater, developed by a localized pressure rise (Perret, 1924)." [The pressure induced by high temperature may be due to gases or liquids.]
Synonym: "tumefaction" (no q.v.), "tumulus," "lava upheaval," "bulge."
See also "tumescence."

invagination
Stamp (1961), p. 263: "Limited local subsidence of a [volcanic structure or its environs]."

*ir   irruption
"Eruption"

is   island volcano
Schieferdecker (1969), p. 238, term 3980: "A volcano which originated submarine or in a large lake, having its top part projecting above the water."
See also "volcanic island."

ju   juvenile ejecta
[Material thrown out of a crater or vent which is consanquineous to the associated lavas.]

ka   kawah
*   Schieferdecker (1959), p. 240, term 4011: "Indonesian word, used for different volcanic manifestations. It may mean a real crater, a bocca, a crater lake and also a solfatara field (Escher, 1920)."

Katmaian type of eruption
Fenner (1937), p. 236–239: "Eruptions that form deposits having the characteristics of the [ash] flow of the Valley of Ten Thousand Smokes."

ki   kipuka
Webster 3 (1967), p. 1246: "Hawaiian 'window'; an area [or 'island'] of older land [frequently vegetated], ranging in size from a few square feet to several square miles surrounded by later lava flows. Kipukas result from either topographic irregularities or the viscosity of the lava (H. T. Stearns and G. A. Macdonald)."
Synonym: "steptoe."

kl   klasmatitic
*   See book review by A. Johannsen of Schneider, (1911). Includes homates, maare.
Synonym: "rheumatitic," "rheuklastitic."

ko   konides
*   See book review by A. Johannsen of Schneider (1911).
See also: "rheumatitic," "rheuklastitic," "klasmatitic."

kr   krakatoan
[Pertaining to a highly explosive volcanic eruption.
Synonym: Krakatauan (preferred).

Krakatau type caldera
Cotton (1944), p. 310: "Williams described Crater Lake as a caldera produced by collapse following the evisceration of a magma chamber by explosions of pumice." All calderas formed in this way are included under the heading of "Krakatau type" from Williams (1942). [A caldera in which explosivity has played a significant to large role].

la labial eruption
Swayne (1956), p. 59: "A fissure eruption."

ladu
"Glowing avalanche" (Dutch).

lahar
Ross and Smith (1961), p. 5: "A term used in Indonesia to designate a volcanic mudflow. The term has been used by Curtis (1954, p. 458) for any volcanic breccia with a matrix of tuffaceous aspect which came to rest as a single unit and was originally mobilized by addition of water, gravity along being the motivating force. Lahars may originate in different ways such as mobilization of rain-soaked debris on volcanic slopes (cold) or by eruption through crater lakes (hot) (van Bemmelen, 1949), p. 191). Hot lahars may also be initiated by nuées ardentes entering streams, [snowfields, or seas]." [Not all mudflows are lahars].

lake
Webster 3 (1967), p. 1265: "A pool of liquid such as lava."

lamination
[The planar array of constituents of lava parallel to the flow surface, or to other laminations.]

lapilli (singular, lapillus)
Ross and Smith. (1961), p. 5: "Wentworth and Williams (1932, p. 33) state, 'According to most writers, lapilli fragments may consist either of juvenile lava, still liquid or plastic when ejected, or of broken rock of any sort from the walls of the vent or from the bed rock; that is, they may be essential, accessory or accidental ejecta.'"
Bayly (1968), p. 40: "Lapilli are pyroclastic fragments with diameters between 4 and 32 mm; thus most agglomerates contain lapilli. However, the term is particularly used for somewhat rounded lumps that form by accretion, as drops of fluid magma and ash fly through the atmosphere together."
Synonym: "lava pellets." Adjectives: "scoriaceous," "porous," "pumiceous."

lapilli ash
Macdonald, G., in Hess, and Poldervaart, (1967), p. 52: "Name of rock where lapilli are sufficiently numerous in ash; either scattered through ash or form more or less discrete beds in ash."
Both lapilli and ash are pyroclastics, each of a specific size:

Dust:	<0.05 mm
Ash:	0.05–4 mm
Lapilli:	4–32 mm
Blocks:	>32 mm

Preferred "lapilli tuff".

lapilli cone
[A cone built up of lapilli-sized pyroclastics.]

lateral cone
[A cone built up on the flanks of a volcanic structure.]

lateral crater
M. Campbell: "A small crater formed on the flank of a large volcanic cone." See "volcanic cone."
Synonym: "parasitic crater," "adventive crater."

lapilli flow
[Fragmental material predominantly in the size range 4 to 32 mm.]
See "block flow."

lateral moraine
[Volcanic debris built up on the sides of a lava or ash flow.]

lapilli-stone
[A volcanic rock composed of fragments from 4 to 32 mm in diameter.]
Synonym: "lapilli tuff."

lapilli tuff
Schieferdecker (1959), p. 259, term 4267: "An indurated deposit, essentially made of lapilli fragments from 4 to 32 mm in diameter in a fine tuff matrix (Wentworth and Williams, 1932). When the fine matrix exceeds that of the lapilli, the rock may be called tuff with many lapilli (Blyth, 1940)."
Synonym: "lapilli stone."

* lapillo
Webster 3 (1967), p. 1271: "Lava in the form of lapilli."

* lauoho o pele
Johannsen (1939), p. 262: "The Hawaiian name for Pele's hair."

lava
Stokes and Judson (1968), p. 509–510: "A general name of molten rock poured out on the surface of the earth by volcanoes and for the same material that has cooled and solidified as

solid rock."

Challinor (1964), p. 141: "The term is often extended to cover the magma, and corresponding rock, which, though not actually extruded, is in direct connection with the outpoured material, such as that in a volcanic peak."

Playfair (1802) also terms the magma of a minor intrusion as 'unerupted lava'."

lava ball
Schieferdecker (1959), p. 258, term 4255: "[A lava object] of spherical or ellipsoidal shape, from a few meters to less than one decimeter in diameter. The crust is generally compact, the center scoriaceous. According to Dana (1890) the shape results from friction of the blocks in a moving lava flow. Perret (1924) found such blocks already formed in the vent." See also "accretionary lava ball."

lava bed
[A unit flow of lava.]

lava blisters
"A more or less circular upwarp in a lava melt or crust. A tumulus is usually a less regular upwarp and can be formed by a flow or post-flow process. A tumulus can be larger than a lava blister formed by a gas bubble."

lava cave
"A cave in lava terrain, usually within a fracture (gja), tumulus, or collapsed lava tube."
Synonym: "lava cavern."

lava cavern
See "lava cave."

* lava cauldron
Synonym for "lava pit."

lava colonnade
[An exposure of columns in a lava flow or cooling unit. Colonnades also occur in tuffs.]

lava column
Schieferdecker (1959), p. 243, term 4051: "Column of fluid or solidified lava in a volcanic vent."
See also "volcanic column."

lava cone
Webster 3 (1967), p. 1279: "A volcanic cone is composed predominantly of lava flows." [Sometimes the cone displays a convex profile due to flank flows of viscous lava.]

lava cupola
Schieferdecker (1959), p. 242, term 4032: "A lava dome."

lava delta
Naval Oceanographic Office, p. 95: "A delta-like body of 'lava' formed where a lava flow enters the sea. A coast consisting of such deltas formed by recent lava flows has a convex shoreline and is called a lava-flow coast."

lava dome
Schieferdecker (1959), p. 242, term 4032: "Protrusion of highly viscous lava forming a steep mound above its vent. Its form is different whether it originates in a crater, on the flank of a volcano or isolated on a fault. Unfortunately it is also used to designate shield volcanoes as at Hawaii, or the tumuli that develop on lava flows." [The term should not be applied to shield volcanoes or tumuli.]
Synonym: (less-preferred) "volcanic dome," "plug dome," "tholoid," "cumulo volcano," "cumulo dome," "lava cupola."

lava eruption
Schieferdecker (1959), p. 263, term 4321: "Volcanic eruption emitting lava flows with or without minor explosive phenomena."

lava field
[A broad area of lava flows, rough or smooth, with or without vent structures, eroded or non-eroded. Some pyroclastic cover may be present.]
Synonym: "lava plain."

lava flood
Schieferdecker (1959), p. 250, term 4148: "Enormous emission of lava, which flows out rapidly with an output of millions of tons hourly, leading to [the accumulation of many] lava sheets [to form a] basalt plateau."

lava flow
Stokes and Varnes (1955), p. 82: "A stream or sheet of lava whether molten or congealed. A lava flow may be subaqueous or subaerial depending on whether its final cooling takes place under water or on dry land." [On other planets, lava may congeal in vacuum, or under heavy gases or ices.]

lava fountain
Schieferdecker (1959), p. 266, term 4374: "A gas-charged fluid [emitted from a lava lake, volcanic crater vent or fissures] shooting up

regularly [or erratically] to heights ranging between a few meters and more than [500] meters above the lake surface."
Synonym: "fire fountain" (H. S. Stearns).

lava-flow plateau
Schieferdecker (1959), p. 112, term 1941: "A plateau formed by extensive, solidified lava flow."
See also "lava plateau."

lava froth
Nelson and Nelson (1967) p. 210: "A highly vesicular lava."
See "pumice," "reticulite scoria."

lava lake
[Molten lava filling or partial filling a crater floor. Nyiragongo is an example. If the lake has solidified, the adjective "frozen" or "solidified" should be added.]

lava levee
Schieferdecker (1959), p. 252, term 4177: "The frozen sides of a lava flow forming a border of irregularly piled-up blocks of lava."
See also "scoria moraine."
Synonym: "lateral moraine" (less-preferred).

lava mesa
[A flat-topped hill capped with a near-horizontal lava flow.]

lava ogive
Stamp (1961), p. 344: "Thorarinsson (1953a) extends the use of the glaciology term 'ogive' to 'low waves on the surface of a lava stream'."

lava pellets
Schieferdecker (1959), p. 256, term 4231: "Lapilli."

lava pit
Stokes and Varnes (1955), p. 109: "Pit crater occupied by a lake of lava (from Billings)."

lava plain
See "lava field" and "lava plateau."

lava plateau
[A regional expanse of great thickness of lava flows. The plateau may be dissected. i.e., Deccan, Columbia River.]

lava plug
Schieferdecker (1959), p. 242, term 4041: "The cooled lava in a volcanic pipe, frozen solid dur-

†

ing the period of dormancy to a depth of a hundred meters or more (R.A. Daly,)."
Synonym: "volcanic plug."

lava rag
Schieferdecker (1959), p. 258, term 4258: "A small mass of scoriaceous lava of indefinite shape, thrown out of the eruption."
See "fiamme."
Synonym: "lava clot," "slag lump," less preferred.

lava residual
Nelson and Nelson (1967), p. 210: "The lava flow remaining after twin laterals have cut deep valleys on either side."

lava ring
Schieferdecker (1959), p. 251, term 4155: "A congealed wall of lava rising above the average level of a lava-filled pit, sometimes built up by [a fluctuating lava lake level, lava waves, subsequent lava drainage or spatter]."

lava river
Schieferdecker (1959), p. 250, term 4138: "A lava flow."

lava scarp
Schieferdecker (1959), p. 251, term 4158: "The slope of the lava flow."
[A faulted erosional or topographic flow-produced cliff or steep face of a lava flow or assemblage of flows.]

lava scratches
Schieferdecker (1959), p. 253, term 4183: "Parallel scratches on lava surfaces produced by sliding of lava flows over each other shortly before consolidation, and indicating the direction of the movement." See also "grooved lava."
Synonym: "grooving marks" (no q.v.).

lava sheath
Schieferdecker (1959), p. 252, term 4170: "The lava lining enveloping the empty tube [of a tree mold]."

lava sheet
[A tabular and areally large (compared with thickness) lava flow.]
Synonym: "extruded sheet."

lava shield
[A convex carapace of lava or assemblage of lava flows applied to submeter to multi-kilo-

meter-sized surfaces or structures.]

lava sink
Stokes and Varnes (1955), p. 82: "The founder-ing or sinking in of part of the lava surface pro-producing vertical-walled pits, called lava sinks, in which molten lava may [or may not] rise or cascade (A. K. Lobeck)."

lava spine
Longwell, et al., 1969, p. 643: "An upright, cylindrical feature created by the upward squeezing of a mass of sluggish, pasty lava." [No jetting or eruptive action is implied as distinct from a lava spire or hornito. Also not a driblet cone.]
Hobbs (1919) "An erect cone of lava at the floor of a lava tube formed by lava dripping from a stalactite above. Preferred term "lava stalagmite."

lava stalactite
Schieferdecker (1959), p. 252, term 4173: "A pendent cone formed by dripping of remelted lava from the roof of an empty lava tunnel (Hobbs, 1919)." Note: Jagger (1947) distin-guished, 'worm stalactite', 'walking-stick stalac-tite', 'grape-vine stalactite'."

lava stalagmite
[A cone usually on a lava tube floor built up from drippings from above. Often under a lava stalactite (q.v.).]
Preferred non-synonymous with "lava spine" (q.v.).

lava stream
[A lava flow or river.]

lava subsidence scarp
Sharpe: "A scarp produced by subsidence in lava terrain."

lava tableland
Schieferdecker (1959), p. 112, term 1941: "A lava-flow plateau" [For a detailed discussion of true Icelandic lava table mountains, see Bem-melen and Rutten 1955.]

lava toes
Schieferdecker (1959), p. 250, term 4141: "Small bulbous protrusions with distinct con-centric arrangements, in the front of an advanc-ing pahoehoe lava, generally [less than a meter] in diameter, and only a few [decimeters] long (Macdonald, 1953),"

lava tongue
Schieferdecker (1959), term 4140: "A short lava flow or a ramification, with a maximum length of 2–3 km."
Less-preferred synonym: "lobe of lava" (Macdonald, 1953).

lava tower
Schieferdecker (1959), p. 251, term 4156: "Ragged pillar, predominantly consisting of scoriaceous material in the core, with a coating of denser lava. These pillars, having diameters of several meters and heights which may attain more than 10 meters, are formed by the scoriae which consolidate at the surface of a lava flow or lake, and concentrate in eddies. After the drainage of the molten lava these scoriaceous accumulations are left behind as pillars (Bem-melen and Rutten, 1955)."

lava tree
Stearns: "A lava tree mold which rises above the level of a lava flow."

lava tree cast
[The rare filling of a lava tree mold that forms casts reproducing the shape of the original tree trunk.]

lava tree mold
Macdonald, G. in Hess and Poldervaart, (1967), vol. 1, p. 35: "Where fluid lava sur-rounds a tree it may be chilled, and a cylinder of partly to wholly solidified lava may be formed. The tree may then burn to charcoal and ash, which in turn may be removed by wind and water, and perhaps by organisms, leaving the cylinder empty. The inner wall of the cylinder may preserve in great detail a mold of the surface of the bark"—p. 36: "They may project as much as 7 or 8 m above lava surface and are sometimes referred to as 'lava trees'. Molds may also be formed around fallen tree trunks and limbs."

lava tube
Schieferdecker (1959), p. 251, term 4165: "The tubular tunnel inside the sheath of frozen lava where the still-fluid lava flows with compara-tively slow further loss of heat. After the supply has stopped and the contents of the upper part of the tunnel have flowed downwards an empty tube is left (Hobbs, 1919)." [Lava tubes may be many kilometers long and may form at the end of a lava flow channel. Collapse of the lava tube roof may lead to a series of depressions in a sinuous alignment.]

Synonym: "lava tunnel," "supply channel" (no q.v.), "feeding tube."

lava tube cave
Halliday (1966), p. 380: "A cave formed by the collapse of a portion of a lava tube. A finite section of an uncollapsed lava tube."

lava type
Tomkieff (1940), p. 113: "[A specific variety] of effusive rock characterized by a definite mineralogical as well as by a definite chemical composition."
Cf. "magma-type."

lava tunnel
See "lava tube."

lava upheaval
See "intumescence (of lava)," (Schieferdecker, p. 242, term 4035).

e   lekolith
Coats (1968), p. 71: "A mass of extrusive igneous rock more or less equant in plan, with a nearly level surface determined by the shape of the basin that it filled, and a diameter greater than its depth"—from Greek lekos, "a basin or dish."

leucolitic fumarole
Schieferdecker (1959), p. 275, term 4489: "According to L. Palmieri about 1907 a fumarole rising from a red-hot lava."
Cf. "croicolitic fumarole."

levee
See "lava levee."

li   linear eruption
See "fissure eruption."

liquid lava
[Molten lava.]

lithic
Stokes and Varnes (1955), p. 85: "An adjective applied to any pyroclastic deposit in which the fragments are composed of previously formed rocks; for instance, accidental pieces of sedimentary rock, the accessory debris of earlier lavas in the same cone or shattered bits of earlier lavas in the same cone, or even shattered bits of new magma that first solidified in the vent and then were blown out (Wentworth and Williams, 1932)."

lithic crystal tuff
See "crystal lithic tuff" and "crystal tuff."

lithic tuff
Ross and Smith (1961), p. 8: "In describing lithic tuffs, Pirsson (1915, p. 201) states that the essential feature of tuffs of this class is the presence in them of a striking or dominating degree, of fragments or previously formed rocks. These fragments may be holocrystalline or partly glassy. Rock fragments are present in nearly all ash-flow tuffs but rarely exceed a few percent of the rock." Hatch and Raskall (1961, p. 322,) also state that "Such tuffs often show the rounded outlines and vesicular structure of lapilli."

littoral cone
Macdonald, G., in Hess and Poldervaart, (1967) vol. 1, p. 56: Cone formed at shoreline of ejected debris where aa lava flows enter the water. It occurs where rootless steam explosions continue for hours or days. Some are as much as 100 m high and more than 1 km in basal diameter. They generally are better bedded than typical cinder cones and consist of vitric ash of sand and silt grades enclosing irregular lapilli and bombs up to [a meter] across. [Such cones are common in Hawaii.] See "terminal cone."

littoral explosion
[Explosions produced by contact of lake, river or sea water with molten lava.]

lo   louderback
Stamp (1961) p. 300 quotes Thornbury, (1954) p. 258: "The term louderback was proposed by Davis (1930) for displaced segments of a lava flow on two sides of a fault. If it can be established that the lava flow is of late geologic age there is justification for assuming that associated scarps were produced by faulting [and not erosion]."
Named after the geologist Louderback who first described the phenomenon. An American term not in general use in Britain as yet (G. T. Warwick).

ma   maar
[A relatively round crater less than 10 kilometers in diameter formed by a single eruption or short-lived eruptions or jetting action. It is

surrounded by low to negligible walls of ejectamenta of either younger volcanic debris and
older surface rocks or both.

Maars often occur in topographic lows, are
often filled with water (on earth) and rarely
contain internal cones. The term was originally
applied to crater-lakes in the Eifel. However,
the term, "maar" should be used for the crater,
not the filling.] See "explosive lake."

macula
Hobbs (1953) p. 356: "A magma macula has
been fused within the shale layer beneath the
capping limestone and the transitional calcareous shale beneath it."
"Local pocket in which lava forms by fusion of
shale."
[Local pockets of magma of fused shale.]

magma chamber
Krauskopf (1967), p. 455: "A restricted accumulation of magma at a shallow depth (a
few hundred meters to a few kilometers)
beneath the volcanic structure into which
magma has moved from an unknown source
below." See "ingisept."
Synonym: "magma reservoir."

magma column
Schieferdecker (1959), p. 230, term 3860: "The
fluid magma filling of the volcanic vent."
Synonym: "magmatic column" (no q.v.).

magma reservoir
See "magma chamber."

magmatic eruption
Schieferdecker (1959), p. 265, term 4355: ["An
eruption originating] in the magma itself and
not being caused by influences from outside.
See also; "phreatomagmatic eruption" and
"volcanic eruption."

magmatic pressure
Schieferdecker (1959), p. 229, term 3854: "The
pressure exerted by a magma on its surroundings; the pushing of hypomagma toward the
surface may be due to the internal gas pressure
or, as thought by others, to an external cause,
like folding or faulting movements (Cotton,
1944)." See "volcanic pressure."

malpais
Stokes and Varnes (1955), p. 88: "A term used
commonly in Spanish-speaking regions for the
rough, blocky terrain on the surface of a lava
flow, especially basalt." Literally, "bad land."

mamelon
Nelson and Nelson (1967), p. 223: "Small,
steep-sided volcanic dome composed of light-
colored lavas such as trachyte and rhyolite and
similar viscous rocks." Early applied to domes
in Reunion (Himus, 1955, p. 145–146); [Any
dome-shaped protuberance.] French: "lava
dome."

marginal explosion crater
[A crater formed by explosive action on the
margins of a volcanic structure.]

marole
Krauskopf (1967), p. 457: "A gas vent, smaller
than the vent of a volcano."
Synonym: "fumarole."

massive eruption
Schieferdecker (1959), p. 236, term 3955: See
"fissure eruption."

me **medial moraine**
[The strip of land common to two lava streams
and made up of debris contributed by both of
these flows. The slower moving debris-laden
center of a wide lava flow.]

megata
Japanese term for "maars."

mi **mixed avalanche**
Schieferdecker (1959), p. 273, term 4452: "An
avalanche consisting of a mixture of (a) incandescent and old rocks of the volcano (Bemmelen 1949) or (b) [a mixture of dust] ash
and coarse materials."

mixed eruption
1. An eruption of new and old material.
(Lacroix, 1936).
2. Explosions accompanied by lava flows.

mix-lava
Walker and Shelhorn (1966), ESR vol. 2, p. 99:
"Mix-lavas composed of rhyolite bearing an
abundance of basalt inclusions, have been
described from the Yellowstone Park and
[Iceland]"; Boyd 1961, p. 403: "mix-lavas
show 'conclusive evidence of the simultaneous
fluidity and mingling of rhyolite and basalt'."

mo **moat**
Webster 3 (1967), p. 1449: "A natural feature

resembling a moat as between a lava dome and the crater wall, often of a crescentic shape (Fenner, 1923) [or around a raised plateau in a caldera]."
Less-preferred synonym: "fosse"

mofette
Monkhouse (1965), p. 205: "A small hole in the earth's surface, from which issues carbon dioxide, with some oxygen and nitrogen, and occasionally water-vapour; e.g. in Auvergne in central France, Java, and Yellowstone National Park, U.S.A. It indicates a late stage in minor volcanic activity."

monogenetic volcano
Stubel (1903): "A volcano formed in one, although sometimes long, eruptive period."

monolith
Schieferdecker (1959), p. 242, term 4042: "A volcanic spine."

monticule
U.S.G.S. Topographic Instructions 5C2, 1961: "A subordinate cone of a volcano (French)."

moraine
[Debris deposited at the margins, centers, or termini of lava or as flows.]

mu   mud balls
Schieferdecker (1969), p. 256, term 4225: "Spherical or somewhat flattened, sometimes silicified mud balls, with a diameter from 2.5 to 85 mm, composed of concentric shells of fine or coarse volcanic ash, formed around a grain of sand or lapilli. They are formed during prolonged rotation in whirling eruption clouds (Stehn, 1932)."
Synonym: "volcanic mud pellets;" [less preferred], "accretionary lapilli."

mudflow
Schieferdecker (1959), p. 270, term 4433: "A torrential water and volcanic mud debris flow of all dimensions. The components do not settle according to size, nor have they rounded edges. Lahars may attain thicknesses of dozens of meters and reach distances up to 40 km." See also "hot mudflow," "cold mudflow," "hot lahar," and "cold lahar."
Synonym: "lahar" (less preferred) and (no q.v.) "torrent of mud" (Geikie, 1902).

*   mud lava

Ross and Smith (1961), p. 5: "The term 'mud lava' has been used to describe some materials in Japan which are not known to be ash-flow materials." Matumoto (1943, p. 30) uses the term "welded mud lava."

multiple-vent basalts
[Basalts originating from more than one vent.]

mud pellet
[A pellet of mud which may be formed by rain falling through a volcanic dust cloud.]

mud-tuff
Fisher (1966) p. 294: Synonym of "dust-tuff."

na   nappe
Monkhouse (1965), p. 210: "In France, the term is used more widely for any overlying, covering sheet; e.g., a lava flow."

ne   neck
Stokes and Varnes (1955), p. 96: "A mass of solidified lava or fragmental igneous rock filling the conduit or vent of an extinct or inactive volcano. . . . They usually appear as conspicuous topographic forms after erosion has removed the surrounding accumulations of tuff and ash."
See "plug," "diatreme."

neo-volcanic
Streckeisen (1967), p. 178: "Young volcanic rocks, of Tertiary and Quaternary age. They usually display a somewhat fresher state of preservation than 'paleo-volcanic' rocks and frequently carry glass in the groundmass."
Synonym: "neo-type" facies of Jung and Brousse, (1959).

nested caldera
[A caldera (q.v.) within a caldera.]

nested cone.
[A cone within a cone.]

nested crater
Schieferdecker (1959), p. 240, term 4013: "A large crater or caldera, containing one or more small craters or boccas in its bottom".
[A crater within a crater.]

nested sink
[A sink within a sink.,]

ne   nested volcano
      [A volcano within a volcano.]

no   noncognate block
      [A block alien in composition and origin
      relative to its matrix.]

      nonluminous detritus
      Schieferdecker (1959), p. 255, term 4211:
      "Ejecta of the same magma and of the same
      eruption, but already cooled down (Perret,
      1924)."
      Cf. "incandescent detritus."

      nonwelded tuff
      Ross and Smith (1961), p. 5: "The term 'non-
      welded tuff' will be applied to those ash flows
      or parts of ash flows that have not become
      welded. The mode of deposition and conse-
      quent cooling of ash-fall materials precludes
      welding, in contrast with ash-flow materials
      which may or may not become welded. Thus
      there should be no confusion, and it seems
      unnecessary to use the term 'nonwelded ash-
      flow' materials each time they are mentioned.
      That is, 'nonwelded tuffs' are always to be
      understood as the corollary of welded tuffs. An
      alternative would be to extend the term 'sillar'
      to include all nonwelded tuffs."

      notch
      Schieferdecker (1959), p. 242, term 4043: "A
      breached opening in the crater rim of a volcanic
      crater."
      [The opening may be erosional or structural or
      artificial in origin.]
      Synonym: "cleft."

nu   nucleation
      Schieferdecker (1959), p. 269, term 4413:
      "Relative to volcanic clouds, the phenomena
      of condensation of vapour from volcanic vents,
      but also from the atmosphere, thus giving the
      illusion of a copious emission from the crater
      (Perret, 1924)."

      nuée
      French term for cloud.

      nuée ardente
      Ross and Smith (1961), p. 5: "Lacroix (1903,
      p. 442–443) proposed the term 'nuée ardente'
      to describe the previously unrecognized type of
      volcanism that characterized the ruptions of
      Soufrière and Pelée in 1902"—p. 6: "The term

'nuée ardente' is well established as the name
for a type of volcanic eruption (see 'Peléan
eruptions') . . . Lacroix clearly recognized the
dual character of the nuée ardente—the overriding dust clouds, and the dense basal part."
"Nuée ardente has come to have two distinct
usages; one for a special type of volcanic eruption, and the other for the clouds that ordinarily
accompany these eruptions—glowing cloud
being the commonly used English equivalent.
Usage has not always differentiated between
the clouds themselves and the dense dust ash-
or block- transporting basal part. If so used,
this basal part would constitute the noncloud
portion of a glowing cloud, which may not even
be glowing. In general, it glows only by reflect-
ing the incandescent underlying ash flow."
[This type of flow is exceedingly mobile traveling over 100 km per hour and is essentially a
gas-glass-dust emulsion.]

ob   obelisk
      "Gr. obeliskos, a pointed pillar."

ol   olivine bomb
      Schieferdecker (1959), p. 257, term 4251: "A
      volcanic bomb consisting of 'a large nodule of
      crystallized olivine, thinly coated with new
      lava'."

op   optical pyrometer
      Webster 3 (1967) p. 1584: "A pyrometer that
      measures temperature by means of determining
      the intensity of the light of a particular wave-
      length emitted by a hot body."

or   orgues
*   Stamp (1961), p. 346: (French) "A mass of
      vertical prisms having the appearance of organ
      pipes, due to the exceptionally rapid cooling of
      a lava flow as at Auvergne."

ou   outbreak lahar
      Schieferdecker (1959), p. 271, term 4439: "[A
      volcanic mudflow which originates when a
      crater lake wall collapses.]"
      See "hot mudflow."

      outburst
      [In volcanology, an eruption.]

ov overflow
Schieferdecker (1959), p. 266, term 4368: "The overflowing of lava from the upper portion of a molten lava column."
See also "summit overflow."

pa pahoehoe
Stokes and Varnes (1955), p. 102: "The Hawaiian word for solidified lava that is characterized by a smooth, billowy, or ropy surface having a skin of glass a fraction of an inch to several inches thick. Pahoehoe is distinguished from the 'aa' type by its smooth surface, and probably also by such internal characteristics as a higher content of glass, and by the gas bubbles being more numerous, more spheroidal, and having smoother walls. Pahoehoe flows often contain large open or partially filled lava tubes which served as under-surface conduits for the advancing lava." [Pahoehoe surfaces may be described as billowy, ropy, corded, draped, festooned or folded.]

palagonite tuff
Johannsen (1939), p. 273: "According to von Waltershausen, a sandstone-like brown tuff containing innumerable angular grains and fragments of a yellow or brown substance, called palagonite, and thought originally to be a definite mineral but now known to be simply devitrified, basaltic glass. The rock also contains fragments of augite and olivine, microlites of plagioclase, and broken pieces of basalt."
See: "basalt globes."

paleovolcanic
Johannsen (1939), p. 290: Rosenbusch (1887 p. 6) used "palaeovulkanisch" for pre-Tertiary extrusive rocks. Cf. "Neo-volcanic." [Usually devitrified porphyries, diabases and melaphyres.] Consult Streckeison (1967, p. 178–179) for petrographic difference from neo-volcanic rocks.

paleovolcanology
[Study of prehistoric volcanic features, rocks, and processes.]

pancake-shaped bomb
[A volcanic bomb of discoid shape.]

parasitic cone
Stokes and Varnes (1955), p. 103: "A small subsidiary cone on the flanks of a larger [volcanic structure]."

parasitic crater
[A small subsidiary crater on the flanks of a volcanic structure.]

parasitic lava flow
Schieferdecker (1959), p. 250, term 4146: "A flowing issuing from a fissure on the slope of or at the foot of a volcanic structure."

paravolcanic
Schieferdecker (1959) p. 220, term 3727: "All secondary manifestations of volcanic phenomena."

paroxysm
"Any sudden and violent action of physical forces occurring in nature, such as the explosive eruption of a volcano or the convulsive "throes" of an earthquake; specif. the most violent and explosive action during a volcanic eruption, sometimes leading to the destruction of the volcano and generally preceded and followed by smaller explosions."
Cf. "catastrophe."

paroxysmal
Webster 3 (1967), p. 164: "of, relating to, or of the nature of a paroxysm, as paroxysmal volcanic eruptions (A. Holmes)."

paroxysmal eruption
Monkhouse (1965), "A volcanic eruption with violent explosive activity, usually after a long period of quiescence of a volcano, during which subterranean pressures accumulate. It is sometimes referred to as a 'Vesuvian' or 'Plinian' eruption after that of Vesuvius in A.D. 79. A paroxysmal eruption of recent times was in 1883, when the island of Krakatau, in the Sunda Straits between Java and Sumatra, blew up."

pe pedionites
*   See book review (by A. Johannsen) of Schneider (1911).
See also: "rheumatitic," "rheuklastitic," "klasmatitic."

peléan
Webster 3 (1967), p. 1667: [Mount Pelée, volcano in Martinique]: "(1) of, relating to, or resembling volcanic eruptions characterized by violent expulsion of clouds or blasts of incandescent volcanic ash; (2) characterized by the extrusion of a viscous silicic spine."

peléan cloud
W. G. Moore: A nuée ardente.

peléan eruption
"A type of volcanic eruption characterized by explosions of extreme violence and the formation of nuées ardentes. The lavas are generally very silicic and very viscous, although some are associated with basaltic volcanoes. Named after Mont Pelée, a volcano on the island of Martinique which in 1902 erupted and destroyed the town of St. Pierre. The eruption of volcano Agung on Bali in 1963 is also considered to be peléan."

pelelith
Webster 3 (1967), p. 1667: "Plug or spine of vesicular or pumiceous lava thrust upward in the throat of a volcano."
[Because of the confusion that may arise between Pelée and Pele, the phrase "volcanic spire" may be more desirable.]

Pele's hair
Named for Pele, the Hawaiian goddess of fire. Macdonald, G., in Hess and Poldervaart, (1967), vol. 1, p. 49: "Natural spun glass consisting of molten lava that solidified in the air as long threads. They may be more than 2 m in length and range from about 0.01 to 0.5 mm in diameter. The finest are formed by very gentle fountaining in a lava lake." [The fibers are blown by the wind from frothy lava or from drops of liquid lava thrown into the air.]
Synonym: "filiform lapilli," "capillary lava."

Pele's tears
Webster 3 (1967), p. 1667: "Small drops of volcanic glass."
Macdonald, G., in Hess and Poldervaart, (1967), vol. 1, p. 49: "Solidified lava drops in the size range of fine lapilli. Many of the tears are of typical drop–shape, but others are cylinders with rounded ends; still others are nearly spherical. They form by fountaining of fluid lava and commonly trail behind them long threads of molten lava that solidify in the air to become natural glass called Pele's hair."

peperino
Holmes (1928), p. 178: "A local Italian name for a soft incoherent yellow-grey tuff, containing broken crystals of feldspar, leucite, biotite, and augite, and numerous rock-fragments embedded in a finely-granular base."
Synonym: "leucite tuff" (Kemp, 6th ed. 1940, p. 287).

pépérite
Macdonald, G., in Hess and Poldervaart, (1967) vol. 1, p. 29: "Pépérite is 'inter-pillow material' consisting of a mixture of sedimentary material and lava fragments, i.e., an intrusion breccia. Irregular tongues of lava intrude the soft, water-saturated sediment, which [fill] the interstices between the pillows and lava breccia fragments, forming an intrusive pépérite (Macdonald, 1939)."

perilith
Schieferdecker (1959), p. 257, term 4250: "A volcanic bomb consisting of a fragment of old rock coated with a skin of congealed new lava (Wentworth-Macdonald, 1953)."
Synonym: "cored bomb."

peripheral vent
Schieferdecker (1959), p. 243, term 4053: "An adventive vent."

ph   phreatic caldera
[An explosion caldera formed by the superheating of near surface water-rich sediments and ash by magma. No lava is erupted.]
Schieferdecker (1959), p. 245, term 4076: According to R. A. Daly, "the caldera of Bandai-san (1888) with a diameter of 3 km and a depth of 650 m, is [an example of] one of these."

phreatic eruption
Schieferdecker (1959), p. 225, term 3792: "[Violent] steam and mud eruption caused by the expansion of volatile matter such as water-sulphur-rich gases, etc., above the roof of a near surface magma chamber igneous body. The ejecta blown out at relatively low temperatures contain no fresh incandescent matter."
Synonym: "semivolcanic eruption," "pseudovolcanic eruption," "secondary eruption."

phreatomagmatic explosion
Macdonald, G., in Hess and Poldervaart, (1967), vol. 1, p. 54: "Produced by rising magma encountering either ocean water or ground water where the water table was close to the surface. Great amounts of steam are formed. This added to the magmatic gases produces violent phreatomagmatic explosions."
[The simple phreatic eruption is not intensified by magmatic gases.]

pi   pillow lava

Macdonald, G., in Hess and Poldervaart, (1967), vol. 1, p. 27: "True pillow lavas, in the stricter sense, consist of a mass of more or less ellipsoidal bodies variously described as pillow-shaped, blister-shaped, mattress-shaped, sack-like, or balloon-like in form. In cross section the 'pillows' are more or less elliptical but they give a general impression of radial structure instead of that of concentric structure characteristic of pahoehoe toes.—p. 28: Basaltic pillow range from 10 cm to 7 m in diameter, but most commonly are less than 1 m in diameter. Clay, sand, or marl, similar to that underlying the flow, often fills the spaces between the pillows and commonly appears to have been squeezed up into the spaces by the weight of the lava masses resting on the unconsolidated water-saturated sediment or ash." Macdonald, G., in Hess and Poldervaart, (1967), vol. 1, p. 26: "Pillow lavas are the result of contact with water or water-saturated materials, or perhaps in rare instances with cooler fluid magma, usually basaltic or spilitic."

pipe
Webster 3 (1967), p. 1721: "3b (3): "The eruptive channel opening into the crater of a volcano; also the filling of such a channel."

pipe amygdule
Schieferdecker (1959), p. 323, term 5171: "The infilling of tubular vesicles (pipes) occasionally found at the base of a lava flow to which they are normal or inclined. The pipes may be due to gases disengaged from underlying moist sediments by the heat of the lava."

pipe vesicle
Macdonald, G., in Hess and Poldervaart, (1967), vol. 1, p. 36: "Well known in lavas of northwest U.S.A. Tubes projecting upward into the base of the flow and ranging from 1–2 cm long to extreme examples of several (p. 37) decimeters length. Their diameter is usually less than 1 cm. Some are single rather uniform cylinders; and others are branched; particularly common are those in which two or more tubes rising from the base of the flow unite upward into a single tube, apparently as the result of the generation of steam from underlying wet surfaces and coalescence of several rising bubbles. See also "vesicle cylinder."

piperno
Ross and Smith (1961), p. 7: "A welded trachy-

tic ash tuff, first described from the Phlegraean Fields in Italy. It is characterized by conspicuous lenses of glass (fiamme). 'Pipernoid' is a phase of the same rock in which a similar structure is revealed under the microscope. Dell'Erba (1892), and later Zambonini (1919, p. 72), concluded that piperno was a tuff, and was deposited at a high temperature."

pipernoidal lava
Schieferdecker (1959), p. 262, term 4307: Clastogenetic lava.

pisolite
[Accretionary structure formed by rain falling through dense volcanic dust and ash clouds. They may grow to pea-size.]
Synonym: "chalazoidite."

pisolitic tuff
[A tuff made up of pisolites.]

pit crater
Stokes and Varnes (1955), p. 109: "Circular volcanic [collapse] depression with steep walls. Typically exposed in the Hawaiian Islands." See also: "lava pit."

piton
Weyl (1966), p. 189: "French word for mountain top, more specifically referring to the silicic endogenous domes and lava spines of Martinique, St. Lucia, etc., in the volcanic Lesser Antilles."

pl   planeze
Monkhouse (1965), p. 239: (French) "A triangular wedge of lava on the slopes of a dissected volcano, narrowing upwards to culminate in a broken projection on a crater-rim. These occur where the lava flows protect an otherwise unresistant cone; e.g., in the Central Massif of France. . . ." [Also applied to a lava capped mesa of the Puy de Dome district where the lava has protected the underlying material from erosion.]

plastic lining
Schieferdecker (1959), p. 230, term 3866: "A layer of viscous, semisolid material, formed when hot lava congeals upon the cooler walls of the vent (Perret, 1924)."

plateau
[A high plain of regional extent.]

plateau basalt
Stamp (1961), p. 367: "[Mafic] lavas, notably basalts, are very fluid at high temperatures and when reaching the surface through fissure eruptions, the lava flows often extend evenly and horizontally over vast areas and build up lava plateaus, especially if there are successive eruptions. Good examples are the Deccan lavas of India (formerly called 'Deccan Traps') and the Columbia-Snake Plateau of the Northwestern United States, each covering over [hundreds of thousands of square kilometers]."
Synonym: "flood basalts."

plateau eruptions
Himus (1955), p. 112 "Eruptions whereby successive sheets of lava covering a wide area, eventually built up a plateau. The magma rises to the surface through numerous parallel fissures, and there is practically no explosive activity; consequently, practically no tuff is produced."

plinian
Schieferdecker (1959), term 4341, "Plinian activity, a modification of Vulcanian activity characterized by a gas-blast eruption of enormous power from an eruptive magma of low or medium viscosity and high gas pressure (Escher, 1945; Bemmelen, 1949), generally taking place after a long period of quiescence. According to von Wolff (1914) an explosive caldera is formed by such an eruption. According to Rittmann (1933) the principal characteristic consists in the fact that during the long period of quiescence, the magma has differentiated in the vent and produced a highly explosive 'head'. A prototype is the eruption of Vesuvius in 79 A.D."

plug
Himus (1955), p. 113: "A roughly cylindrical mass of igneous rock occupying the vent of an extinct volcano, commonly exposed by denudation."

plug dome
Williams (1932): "A volcanic dome characterized by an upheaved, consolidated conduit filling."

plug-volcano
Monkhouse (1965), p. 242: "A mass of viscous acid lava squeezed out of a vent in a molten form, and solidifying as dacite, rhyolite, or trachyte."

plutonic cognate ejecta
[Deep-seated material ejected from vents which are consanquineous with the bulk of the ejecta.]

po **polygene**
Webster 3 (1967), p. 1758: "Originating or developing in two or more ways or at two or more times as a volcano built by a succession of eruptions."
Synonym: "polygenetic."

polygenetic volcano
Schieferdecker (1959), p. 239, term 3997: "A volcano formed by several volcanic outbursts (Stubel, 1903)."

ponded lava
[Lava which accumulates in a large puddle or pond.]

postvolcanic
[Occurring after the main volcanic activity.]

pr **pressure dome**
[Dome produced by subsurface pressure.]

pressure ridge
Macdonald, G., in Hess and Poldervaart, (1967), vol. 1, p. 32: "Elongate upheavals on surfaces of pahoehoe flows (cf. 'tumulus' p. 33). They may reach heights of 15 m or more, but in Hawaii few exceed 5 m. In length, they range up to more than 500 m but most are less than 100 m. Pressure ridges represent irregular jumbles of blocks of flow crust and result from "the buckling of the flow crust near the edge of the flow, where its advance is blocked by some barrier, as a result of the continued nearly horizontal thrust of the moving flow behind them."

primary fumarole
Schieferdecker (1959), p. 274, term 4471: "Formed over a fissure in or on a volcano and fed directly from the main source of activity, giving thus a true index of internal conditions (Perret, 1924)." Cf. "secondary fumarole," "tertiary fumarole."

principal vent
Schieferdecker (1959), p. 243, term 4052: (1) The most important vent in a volcano having multiple summit craters of (2) according to Daly (1914), a vent emanating from an abysally injected batholith." [Definition 1 is preferred.]
Synonym: "main vent" (no q.v.).

principal volcano

Schieferdecker (1959), p. 239, term 3993: "The main volcano in a volcano complex, but according to R. A. Daly a volcano directly fed from an abyssal fissure, i.e., connected with a large intrusive magma."

Synonym: "major volcano," "main cone" (no q.v.).

protrusion

Schieferdecker (1959), p. 266, term 4364: "The lifting of a plug, spine [or dome] by the pressure of rising magma."

Less-preferred synonym: "upheaval (of plug)."

ps   pseudobocca

Schieferdecker (1959), p. 243, term 4054: "From Mercalli (1907): A bocca on a superficially congealed lava flow from which a 'secondary' lava flow issues, i.e., the mouth of a lava tunnel from which the lava flows out freely (A. Rittmann)."

pseudobomb

Schieferdecker (1959), p. 258, term 4255: "A lava ball."

pseudomoraine

Schieferdecker (1959), p. 252, term 4175: "A scoria moraine of a lava flow."

pseudo shield volcano

Schieferdecker (1959), p. 237, term 3965: "A stratovolcano shaped like a shield volcano."

pseudovolcanic

Stamp, p. 377 quotes Thornbury, 1954, p. 515: "Pertaining to certain topographic features [which resemble but are genetically different] from volcanic forms. They include bomb and meteor craters, Carolina bays, salt plugs, etc."

pseudovolcano

"1) An eruptive vent that is not emitting lava like a true volcano.

2) A large crater or circular hollow believed not to be associated with recent volcanic activity; e.g., a crater that is possibly meteoritic in origin but is more probably the result of phreatic explosion or cauldron subsidence." (after Calkins, A.G.I.)

pu   pumice

[Rough cinder-like, light-colored, more or less vesicular lava (pyroclasts) thrown out of an explosive eruption or appearing on a lava stream. The expansion and escape of enclosed gases produce the typical structure. The term is usually restricted to rhyolitic or closely allied lavas. Pumice fragments 4–32 mm are called lapilli.]

pumice cone

[A cone built up of pumice.]

pumice fall

Ross and Smith (1961), p. 7: "Deposition of pumice directly from the air." See "ash fall" and "tephra."

pumice flow

Ross and Smith (1961), p. 7: "Flows with a conspicuous proportion of pumice fragments of lapilli size (4 to 32 mm) or block size (>32 mm) have been called pumice flows. (See "pyroclastic materials.") Tsuya (1930) applied the term 'pumice flows' to the materials formed by the 1929 eruption of the Komagatake volcano in Hokkaido, Japan. Other Japanese geologists, including Kozu (1934, p. 136) and Kuno (1941, p. 148), have described pumice flows. As used by the Japanese geologists, 'pumice flow' is synonymous with 'ash flow'."

pumice tuff

[Tuff made up of pumice fragments.]

pumiceous bomb

Schieferdecker (1959), p. 258, term 4260: "Irregularly elongate pieces of pumice with glassy bread-crust skin."

pumiceous lava

Schieferdecker (1959), p. 249, term 4127: "Foamy lava."

pumicite

Webster 3 (1967), p. 1841: "A volcanic dust that is similar in composition to pumice and used for abrasive purposes."

pumilith

Hess: "Lithified volcanic ash."
See also "pumicite."

puy

"A term used in central France for any isolated hill or mountain; specif. a small volcanic cone in the Auvergne district, representing a remnant of an extinct volcano [or plug dome] of moderate dimensions, and consisting of ash, cinder, or acidic lava." Etymol: Latin *podium* "balcony."

pyroclast
Webster 3 (1967), p. 1854: "A [previously molten] fragment of volcanic material that has been expelled from a vent into [an atmosphere or a vacuum]."

pyroclastic
Stokes and Varnes (1955), p. 116: "Produced by explosive [fragmentation and] ejection of material from a volcanic vent [through an atmosphere or a vacuum] (Wentworth and Williams 1932). Applies to the deposits as well as the textures so formed."

pyroclastic breccia
"Explosion breccia."

pyroclastic cone
A. Poldervaart: (1967) "Not defined, synonymous with ash cone and tuff cone."

pyroclastic debris
See "pyroclastic deposits."

pyroclastic conglomerate
See "pyroclastic deposits."

pyroclastic deposits
Stokes and Varnes (1955), p. 116: (1) "Air [or vacuum fall] deposits made up mainly of rock material that has been expelled from a volcanic vent, such as agglomerate, tuff, and ash. The fragments range in size from great blocks to the finest dust or ash." (2) (p. 320) "True sediments consisting essentially of volcanic ejectamenta, the fragmental nature of the material being due to explosive eruption" [and the texture modified by sedimentary processes, the formation of conglomerates for example].

pyroclastic flow
Fisher (1966), p. 290, quotes Smith in Aramaki and Yamasaki (1963): "All fragmental flows or avalanches composed of pyroclastic material irrespective of temperature of emplacement." For a great number of terms applied to rock produced by pyroclartic flows, see: Pantó (1963), pp. 174–181.
See: "ash flow," "nuée ardente."

pyroclastic material
Macdonald, G., in Hess and Poldervaart, (1967), vol. 1, p. 45: "Pyroclastic material is defined as all the material projected in the air [or at altitude] from volcanic vents. Although pyroclastic material is fragmental, . . . the term is not synonymous with fragmental, because

the tops of aa and block lava flows are also fragmental. Probably the projection always involves some degree of explosion, but in some instances as in the formation of Pele's hair, the explosion may be very slight. As a welcome synonym for the rather ponderous phrase, 'pyroclastic material', Thorarinsson (1953b, p. 6) has revived the term "tephra" used by Aristotle."

pyroclastics
Synonym: "pyroclastic rocks"

pyroclasts
Stamp (1961), p. 378 quotes Holmes (1944), p. 443: "In addition to the eruption of hot gases and molten lavas from volcanoes, vast quantities of fragmental materials are often produced by the explosion of rapidly liberated gases. These materials, collectively known as pyroclasts may of themselves consist of molten or consolidating lava, ranging from the finest comminuted particles to masses of scoriae and volcanic bombs of considerable size, or they may be fragments of older rocks ranging from dust to large ejected blocks. . . ."

pyroclastic sandstone
Fisher (1966), ESR vol. 1, p. 289: "Synonymous with 'tuff' in that it is 'a rock composed of sand-size pyroclastic fragments'. Thus, the term is unnecessary."

pyroclastic texture
Longwell, et al., (1969), p. 657: "A texture of sediments and sedimentary rocks resulting from physical transport and deposition of particles formed by volcanic activity."

pyrogeology
[Volcanology or "fire" geology]

pyrometer
Webster 3 (1967), p. 1854: "An instrument for measuring temperatures (as beyond the range of thermometers) usually by the increase of electric resistance in a metal when heated, or by the generation of electric current by a thermocouple when acted upon by direct intensity of light radiated by an incandescent body as its temperature increases."
See: "optical pyrometer," "radiation pyrometer."

pyrometric cone
Webster 3 (1967), p. 1854: "Any of a series of small cones of different substances that soften and arch over successively as the temperature

rises, that together form a scale of fusing points and that are used in finding approximately the temperature (as of a kiln)," Called also "Seger cones."

pyrometry
Webster 3(1967), p. 1854: "The technique and method of measuring high temperatures; esp. the art of using a pyrometer."

pyrophotometer
Webster 3 (1967), p. 1854: "An optical pyrometer in which light from an incandescent body whose temperature is to be measured is passed through ruby glass and the red rays thus isolated are compared with those similarly received from a standard flame."
Cf. "pyrometer."

* pyrosphere
Webster 3 (1967), p. 1854: "A hypothetical spherical zone of molten magma that is held to intervene between the crust of the earth and a solid nucleus and to supply lava to volcanoes."

qu   quiescence
Webster 3 (1967), p. 1865: "The state of inactivity or repose: tranquility, at rest, motionless, quiet."

quiet
Webster 3 (1967), p. 1865: "Said of a volcanic eruption, marked by the extrusion of lava without violent explosions."

ra   rain lahar
Schieferdecker (1959), p. 271, term 4437: "[A deposit] formed by heavy rains, falling on the slopes of a volcano, and removing loose materials. They may either be hot or cold, depending on the time of origin, i.e., soon after the eruption, or at a later date, when the material has cooled down."

rampart
Schieferdecker (1959), p. 140, term 4406:
1) "A ring-shaped ridge of debris."
2) "Ring-shaped top part of a volcano, as Monte Somma, generally more or less crescentic in shape."
See: "spatter rampart."

rampart crater
[A volcanic crater containing within it a ring or partial ring comprising an earlier or later wall segment.]

re   relaxation
Schieferdecker (1959), p. 267, term 4389: "The remission of pressure in the magma (e.g., after the initial explosions)."

repose
Webster 3 (1937), p. 1926: "Quiescence, as in 'the volcano was in repose'."

repose period
Schieferdecker (1959), p. 267:
1) term 4380: Perret (1924): "A quiet phase of the volcanic cycle, as distinguished at Vesuvius."
2) term 4386: "The time of solfataric (fumarolic) activity between two volcanic outbursts."
Less-preferred synonym: "repose phase" (no q.v.), and "dormancy."
See: "solfatara stage" and "fumarole stage."

reticulite
"An extremely attenuate [and brittle] pyroclastic rock consisting of glass threads that join a series of points in a polyhedral space lattice. It is formed by the collapse of the walls of adjacent vesicles in pumice and the retraction of the liquid into threads which outline the perimeters of the former polygonal faces of the pumice fragments. The threads are usuall of triangular cross-section, indicating that they were chilled too quickly to become rounded. Such rock has generally been known by Dana's term 'thread-lace scoria'."

rh   rheuklastitic
*   See book review (by Johannsen, 1939) of Schneider, (1911). See also: "rheumatitic," "konides," "klasmatitic."

*   rheumatitic
See book review (by Johannsen, 1939) of Schneider, (1911). See also: "pedionites," "aspites," "tholoides," "rheuklastitic," "klasmatitic."

rheo-ignimbrite
Cook (1966), p. 165: "Ignimbrite that develops secondary flowage because of its high temperature and because it comes to rest on a slope."
Rittmann (1958), pp. 524–533: "A pseudopluton representing an erosional remnant of former crater vents of ignimbritic eruptions, formed upon termination of eruption in which the fluidized mass remaining in the crater

vent would settle, de-gas, and fuse, and then crystallize as a holocrystalline, plutonic-like rock. The material fills volcanic vents of magma chambers, or both."

ri   ribbon bomb
Rock (1915): "An elongated, flat and twisted piece of lava."

rim
Schieferdecker (1959), p. 241, term 4027: "Of a crater, the curved border or margin, surrounding the top part of the crater."
Less-preferred synonym (no q.v.): "lip," "brim."
See also: "caldera rim."

rim volcano
[A volcano situated on the rim of a caldera apparently localized by structural control. The eruptions tend to obscure this control.]

ro   roaring fumarole
[A vigorously outgassing fumarole producing more than normal noise.] French synonym: "soffine."

root
[Volcanic source of hot springs, volcanoes, fumaroles, etc.] Schieferdecker, (1959), p. 274, term 4472: "A secondary fumarole." [One not over a primary vent.]

rootless vent
[A vent over a secondary fracture not connected with the source magmas giving rise to the eruption.]

rootless lava flow
Schieferdecker (1959), p. 262, term 4306: "Clastogenetic lava flow."

ropy lava
Schieferdecker (1959), p. 248, term 4122: "Rather fluid lava. Its surface features much resemble ropes coiled on the deck of a vessel. The coils consist of lines of glassy scoriae so arranged by surface currents in the lava stream (Hobbs, 1919). It is a variation of pahoehoe lava. (p. 249) (Hawaii) or helluhraun (Iceland)."
Less-preferred synonym: "corded lava."

rotational bomb
Schieferdecker (1959), p. 257, term 4240: "A volcanic bomb showing spheroidal or closely similar forms resulting from rotation during flight." The term includes: "spindle-shaped bomb," "spheroidal bomb," and "tear-shaped bomb."

ru   rubble
"Chaotic or poorly-sorted rock to dust debris."

sa   sand flow
Ross and Smith, (1961), p. 7: "Griggs (1922, p. 253–254) describes the 'great hot sand flow' as: 'The bulk of the deposit is composed of fine fragments, many of them dust-like, but there are included numerous lumps of punice which, in places, make up a considerable fraction of the whole. There is no trace of stratification of the materials, except where they were obviously subject to secondary readjustments after deposition. At the foot of the valley we found clear-cut and positive evidence of the manner of the tuff, making it certain that the sand of which it was composed must have flowed down the valley like a viscous liquid."

sand sea
Schieferdecker (1959), p. 246, term 4089: "The more or less flat plain formed by ash and other material ejected from central cone(s) and washed by rain over the floor of the caldera."

sand-tuff
Fisher (1966), p. 294: Modifying Bailey (1926): "A pyroclastic rock (showing no effect of erosion) in which component fragments are greater than 50 percent of rock and diameter ranges from 1/16 to 1 mm."
Cf. "tuffaceous sandstone."

Santorin earth
Webster 3 (1967) p. 2012: (Santorin, Greek island in the Aegean): "A volcanic tuff from the island of Santorin consisting principally of fine light gray siliceous material, used for making cement."
Synonym: "pozzolana."

sc   Schollendome
Schieferdecker, (1959), term 4035: A German term for "lava intumescence" or "tumulus."

Schollenlava
Holmes (1928), p. 253: A German term for "block lava."

scorched area
Schieferdecker (1959), p. 249, term 4137: "Part

of the landscape, destroyed and barren, owing (p. 250) to covering by a lava flow."
Less-preferred: synonym "malpais"

scoria cone
[A cone built up of scoria.]

scoria
Stokes and Varnes, (1955), p. 129: "Rough, cinderlike, dark colored, more or less vesicular lava (pyroclasts) thrown out by an explosive eruption or appearing on a lava stream. The expansion and escape of enclosed gases produce the typical structure. The term is usually restricted to basaltic or closely allied lavas."
[Scoria fragments 4-32 mm are called lapilli.]
"Plural is scoriae or scorias."

scoria moraine
Schieferdecker (1959), p. 252, term 4175: "Formed by debris of frozen lava of the crust of the lava stream, falling in front and along the sides of the flow." [Terminal debris of scoria of a lava flow.]
See also: "end moraine", "lava levee", "medial moraine", and "basal moraine."
Less-preferred synonym: "pseudomoraine"

scoria tuff
[A fine grained volcanic rock made up of scoria fragments.]

scorify
[To render frothy, as scoria.]

se   secondary eruption
Schieferdecker (1959), p. 225, term 3792: "A phreatic eruption."

secondary fumarole
Schieferdecker (1959), p. 274, term 4472: "A gas vent developed upon lava flows at a distance from their source; or upon glowing cloud deposits and hot lahars (Perret, 1924)."
Synonym: "rootless fumarole." Cf. "primary fumarole," "tertiary fumarole."

sector-graben
[A downdropped portion of a caldera or volcano rim or flank or floor.]

sedimentary ash
[Any deposit of ash whether transported by air or water].

sedimentary tuff

[Any deposit of tuff whether deposited by air or water.]

Seger cone
Webster 3 (1967), p. 2056: "(after Hermann August Seger, German ceramist)" "Pyrometric cone."

sh   shard
Webster 3, (1967), p. 2087: "Highly angular, curved vesiculate glass fragments of pyroclasts."

sharkskin pahoehoe
"A type of pahoehoe, the surface of which is covered with innumerable tiny spicules and spines produced by the escape of gas bubbles each of which dragged with it a filament of the enclosing liquid lava."

shark-tooth projection
Macdonald, G. in Hess and Poldervaart, (1967), vol. 1, p. 35: "Where one mass of still-plastic lava is pulled away from another, particularly along the edges of lava rivers where the crust tends to adhere to the bank only to be pulled away by movement of the river, and along slump scarps, the lava commonly is drawn out into projections tapering to fine sharp points. These are known as shark-tooth projections. They are commonly less than 5 cm long, although observed up to 25 cm. The tips of the projections point away from the direction of relative movement of the lava mass to which they are attached."

shelly pahoehoe
Macdonald, G. in Hess and Poldervaart (1967), vol. 1, p. 13: "A gas-rich lava, near vents, that develops a large number of small open tubes and toes covered by a crust only 1 to 30 cm thick (Jones, 1943, p. 265–268)."

shield basalts
"Basalts from shield volcanoes."

shield cluster
See: "volcanic shield cluster."

shield cone
Webster 3 (1967), p. 2094: "A conical or domical shield volcano."
Synonym: "shield dome."

shield volcano
Stokes and Judson (1968), p. 517: "A [relatively low, broad volcano, many kilometers in cir-

cumference] built up almost entirely of lava, with slopes seldom as great at 10° at the summit and 2° at the base, [producing a shield-like profile]." [Little to no fragmental material is present.] Examples: Mauna Kea on the island of Hawaii.

si   sillar
Hatch and Rastall (1965), p. 322: "An ignimbrite that lithifies after deposition by recrystallization due to the activities of escaping hot gases and fluids." Ross and Smith, (1961), p. 5: Term proposed by Fenner (1948, p. 883): "One of principal characteristics of the poorly indurated tuffs of the Arequipa region of Peru is induration by 'vapor-phase crystallization'." Ross and Smith feel that "sillar" could be "more appropriately used for tuffs indurated by this process than as a general term for non-welded tuffs". Sillar has been used synonymously for "nonwelded ash-flow tuffs."

simple caldera
Schieferdecker (1959), p. 246, term 4081: "Formed by one great outburst only. Spethmann used this term for a caldera formed by a single volcanic phenomenon in distinction to a 'kombinierte Kaldera', formed by concerted action of [many volcanic processes]." [These are considered to be a rare type of caldera, most being polygenetic.]

simple volcano
Schieferdecker (1959), p. 238, term 3988: "A regularly built volcano with a single [more or less permanently located summit pit]."

sink
[In volcanology, a simple depression formed by collapse.]
See: "pit crater."

sinkhole
Monkhouse (1965) p. 282: "A volcanic sinkhole [may be produced] in a large shield volcano or lava dome [by withdrawal of molten lava]. Subsidence may form a large depression, several hundreds of meters deep, and over a kilometer across."

sk   skin lava
Schieferdecker (1959), p. 249, term 4125: "Fluent lava."

sl   slab pahoehoe

Macdonald, G. in Hess and Poldervaart, (1967), vol. 1, p. 14: "Pahoehoe flow in which movement of underlying liquid results from breaking of the crust into a series of slabs, a fraction of a meter to 2 m or more across, and generally 5 to 25 cm thick. The slabs may be irregularly tilted and jumbled, or fairly regularly imbricated. It appears to result from an increase in the viscosity of the liquid flowing beneath the crust, with resultant increase in the amount of drag on the crust."

slag
Webster 3 (1967), p. 2137: "Scoriaceous lava from a volcano."
See also: "volcanic slag."

slag bomb
Schieferdecker (1959) p. 258, term 4259: "A bomb of very scoriaceous [texture] (von Wolff, 1914)."

slag cone
Schieferdecker, (1959), p. 240, term 4005: "A cinder cone."

slaggy
Geikie (1897): "When a lava presents an irregularly vesicular character, like that of the slags of an iron-furnace, it is said to be 'slaggy'."
See "scoria."

slump scarp
"A low cliff or rim of thin solidified lava occurring along the margins of a lava flow and against the valley walls or around steptoes after the central part of the lava crust collapsed due to outflow of still molten underlying layers; the inward-facing cliff may be several meters high. Term introduced by Finch (1933), but Sharpe and Stewart (1938, p. 70) would prefer 'lava subsidence scarp' or 'lava slump scarp'."

sm   smoke
1) Steam from fumarole
2) [Aerosol] laden fluids (often dark colored) emitted from volcanic vents.]

smoke-hole
Stamp (1961) p. 202: "Fumarole."

so   soffione
Webster 3 (1967), p. 2165: "A jet of steam usually accompanied by other vapors that issue from the ground in a volcanic region."

Italian, from "soffio," puff, blast, plural: "soffioni." [Some reserve the term to 'steam fumaroles' emitting boric acid vapors.]
Synonym: "fumarole."

solfatara
Monkhouse (1965), p. 287: (Italian) "All vents emitting sulphurous gases, SO₂, H₂S, usually associated with the approaching extinction of volcanic activity. It is so called after a small volcano of the same name in the Phlegraean Fields, near Naples."
Cf. "fumarole," which issues mainly steam and water-vapour.

solfatara field
Schieferdecker (1959), p. 276, term 4499: "A more or less large field of solfataras [disposed] in rows on fissures or apparently in [a random fashion,]."
Less-preferred synonym: "solfataric area" (no. q.v.).

solfataric
Stokes and Varnes (1955), p. 139: "1. Of or pertaining to a solfatara or its action. 2. Pertaining to the stage of feeble activity in a volcanic cycle characterized by emanations of steam and other gases [usually sulfur bearing]."

solfatara stage
Schieferdecker (1959), p. 276, term 4500: "(1) the condition between two volcanic outbursts (see 'repose period'). (2) The condition of a volcano which no longer has any magmatic eruptions though not yet being completely extinct. Sulfur bearing gases are emitted."
See: "fumarole stage."
Less-preferred synonym: "solfatara condition;" "solfataric state of activity" (no. q.v.).

solfatara mound
[A mound built up by solfataric activity.]
Cf. "fumarole mound."

solfataric phase
See: "solfataric stage."

somma
Stamp (1961), p. 426: "(Italian) Originally the rampart remaining from the old crater of Vesuvius and forming an arc around one side of the new one. The name is sometimes extended similar formations in other volcanoes. . . . The derivation from Monte Somma has been misunderstood by some writers and the term has been applied, quite wrongly, to small subsidiary

craters inside a caldera, almost the exact reverse of the proper meaning. See Inter. Geog. Union, 1958, Regional Conference in Japan, Report, 1959."

somma crater
Schieferdecker (1959), p. 239, term 3992: "A crater defined by a more or less complete ring-wall surrounding a central cone."

somma ring
Schieferdecker (1959), p. 241, term 4022: "A large and old, more or less complete crater rampart, surrounding a central cone, as found at Vesuvius (Cotton, 1944)."

somma volcano
[A volcano containing a volcanic cone. A volcano cannot contain a caldera.]
Synonym: "cone-in-cone structure."
Cf. "nested volcano."

soufrière
[French synonym of solfatara.]

sp   spatter
Macdonald, G, in Hess and Poldervaart, (1967), vol. 1, p. 47: "Accumulation of cow-dung bombs and lapilli."
See also "agglutinate driblet."

spatter cone
1) Monkhouse (1965), p. 288: "A large mass of lava ejected violently from a volcano, which 'spatters and congeals as it hits the ground to form a small cone, 3–6 meters high."
2) Schieferdecker (1959), p. 253, term 4180: "A small chimney built up on a lava flow, and consisting of clots of lava, dragged along by the escaping gases, and deposited around the opening of emission (Escher, 1929)."
Synonym: "hornito, "driblet cone," "blowing cone," spiracle;" (less correct syn.) "lava cone" and "agglutinate cone."

spatter rampart
Macdonald, G., in Hess and Poldervaart, (1967) vol. 1, p. 53: "Long low ridge of spatter built along the fissure by fountain chains in early stages of eruption. May be several km long, although generally they are not continous over the entire distance. They may be 5 or 6 m high."

spheroidal bomb
Macdonald G., in Hess and Poldervaart (1967),

vol. 1, p. 47: "A bomb in which the liquid blebs are sufficiently fluid so that surface tension draws them into shapes approaching spheres."
Synonym: "spheroidal bomb," "globular bomb."

spheroidal bomb
See "spherical bomb."

spindle-shaped bomb
Schieferdecker (1959), p. 257, term 4244: "A bomb spun into a spindle shape by rapid rotation during its passage through the air and generally flattened as a result of impact with the ground (Cotton, 1944)."
Less-preferred synonym: "almond-shaped bomb," "bipolar fusiform bomb" (no q.v.).

spine
Webster 3 (1967), p. 2196: "A pointed mass of viscous or solidified lava that occasionally protrudes from the throat of a volcano. (Bayly, p. 36) as at Mont Pelée, where the spine was 300 m high."
Stokes and Varnes (1955), p. 140: "Any pointed projecting eminence is a spine; spines associated with volcanic rocks may range in length from a fraction of an inch to many hundreds of feet (M. Billings)."

spiracle
Macdonald, G., in Hess and Poldervaart, (1967), vol. 1, p. 38: "An opening or vent formed by explosive disruption of the still-fluid lava by gas generated below it. Usually less than a meter in diameter and less than 5 m high. They are common in lavas of northwest U.S.A. where gas jet heights reach at least 12 m with diameters of 10 or more. The jet may contain non-volcanic material blasted up into them by the exploding steam."

spire
[A type of hornito characterized by extreme height relative to diameter.]

spongy
Schieferdecker (1959), p. 332, term 5202: "A term applied to a vesicular rock structure with thin partitions between the vesicles and thus resembling a sponge."

sprout (used pl)
Schieferdecker (1959), p. 249, term 4123: "Protuberances causing the rough, jagged or spinous surface of aaa lava. They are very irregular in form and range from a few milli-

meters to more than 20 cm. (Macdonald, 1953)."
Less-preferred synonym: "spinose sprouts" (no q.v.).

sq   squeeze-up
Macdonald, G., in Hess and Poldervaart, (1967), vol. 1, p. 34: "Two kinds occur, bulbous and linear. Fluid lava from the central part of a flow, either of pahoehoe or aa, commonly is injected into fractures in the solidifying crust, and many of these auto-intrusive dikes reach the surface several decimeters across and several decimeters high." [They are often fin-shaped in profile.]

st   stalagmite
[A positive structure in the floor of a lava tube or under a spatter cone wall overhang where dripping lava builds up a mound.]

stalactite
Stokes and Varnes, (155), p. 141: "A structure (similar to carbonate) developed by freezing; as a stalactite of lava or ice."
See also: "lava stalactite."

steam fumarole
Schieferdecker (1959), p. 275, term 4491: "A fumarole emitting water vapour and containing different volcanic gases."
Synonym: "steam vent."

steptoe
Stamp (1961), p. 631, quotes Knox, 1904: "Island-like areas in a sea of lava." Stamp, (1961), p. 431, quotes von Engeln, (1942), p. 592: "The lava, when emitted, was very fluid ...; keeping its appropriate level, the lava wraps around the spurs of mountains and extends between them as bays. Isolated summits were converted to islands, called steptoes, in the seas of molten rock." "An island-like area of older land surrounded by later lava flows (R. G. Schmidt called 'dagala' in the Etna region and 'kipuka' in Hawaii (Perret, 1924)." French term is 'nunatak volcanique."
See "volcanic nunatak."

stony rises
Stamp (1961), p. 431: "In Australia extensive stretches of lava are often characterized by ridges or stony rises separated by valleys [filled by] flow of the lava when liquid. Isolated blocks or lava blisters may also occur (E. S.

Hills)".

stratified cone
[A cone made up of stratified pyroclastic or flow layers.]

stratified dry tuff
Schieferdecker (1959), p. 259, term 4274: "Subaerial tuff formed by sorting of the particles in the air, the heavy particles falling first."

stratovolcano
"A volcano composed of explosively erupted cinders and ash with occasional lava flows as contrasted with a shield volcano."
Synonym: "stratified volcano" (Cotton, 1944), "composite cone," "bedded volcano," "cone of mixed type" (no q.v.), "Vesuvian-type volcano."

stream
Webster 3 (1967), p. 2258: "A lava flow especially long and narrow."

strombolian
Webster 3 (1967), p. 2265: "(Stromboli, volcano in the Lipari islands): Relating to volcanic eruptions that explode violently and eject incandescent dust, scoria, and bombs with little water vapor."

stufa
Stokes and Varnes (1955), p. 145: "A jet of steam issuing from a fissure in the earth (Merriam-Webster)."

su   subactive volcano
Schieferdecker (1959), p. 224, term 3780: "A volcano which has been active in recent times but of which no outbursts are known (Geze, 1943)."
See "active volcano."

subcrater
Schieferdecker (1959), p. 241, term 4024: Synonym: "Adventive crater (Gregg, 1956)."

sublimate
[A solid deposited from a gas or vapor, such as sulfur, sal ammoniac, hematite, cotunnite.]

submarine explosion
[A volcanic explosion which occurs beneath the sea.]
See "hydroexplosion."

submarine flood tuff

"Marine Ignimbrite."

submarine volcano
Schieferdecker (1959), p. 238, term 3978: "A volcano which erupts and evolves under water."

subordinate volcano
Schieferdecker (1959), p. 239, term 3994: "Derived from an intrusive magma at shallow depth, as sills or small laccoliths (Daly 1914). This term is also used for parasitic volcano." See "parasitic cone."

subsidence caldera
Type of caldera recognized by Thornbury (1954, p. 501).
See "caldera." Cf. "explosion caldera."

subterminal crater
Schieferdecker (1959), p. 241, term 4021: "Situated in the outer slope of the volcano but near the top (Bullard, 1953)."

subterminal outflow
Schieferdecker (1959), p. 266, term 4368: "Lava appearing on the slope of the volcano just below the top."

subaqueous tuff
Schieferdecker (1959), p. 259, term 4276: "Tuff formed by subaquatic eruptions. Near the eruption point the tuffs are nonstratified, at greater distances stratified."

sulfur balls
Schieferdecker (1959), p. 256, term 4226: "Sulphur (mud) balls are small spherical skins of sulphur mud enveloping bubbles of hot volcanic gases, congealed in the open air. They have a diameter from 1 to 4 mm (Stehn, 1932)."

sulfur mud
Schieferdecker (1959), p. 276, term 4507: "A mixture of dark-coloured rock dust mixed or coated with yellow sulphur."

sulfur-mud column
Schieferdecker (1959), p. 277, term 4516: "A surface column of sulphur mud sticking together. Lateral openings in the column permit growth in various directions whereby capricious ramifications of sulphur are formed. The columns at Papandajan (Java) reached heights of 2 meters (Stehn, 1928)."
Synonym: (no q.v.) "sulphur-mud tree."

sulfur-mud flow
Schieferdecker (1959), p. 277, term 4515: "[A flow] issuing from a mud volcano or mud pool. At Papandajan (Java) a length was reached of 150 m with a width of a few decimeters (Stehn, 1928)."

summit caldera
[A caldera formed at the summit of a large volcanic complex or province.]

summit crater
Schieferdecker (1959), p. 241, term 4020: "A crater situated in the top of a volcano."

summit eruption
Schieferdecker (1959), p. 265, term 4357: "Taking place in the top of a volcano, either explosively or effusively."
Synonym: (no q.v.) "top eruption."

summit overflow
Schieferdecker (1959), p. 266, term 4367: "Lava flowing from a crater and over its rim (Dana, 1890)."
Less-preferred synonym: "superfluent lava flow."

sunken caldera
[A caldera of subsidence.]

superfluent lava flow
See "summit overflow."

supravolcano
Schieferdecker (1959), p. 222, term 3748: "A volcanic body forming part of the landscape."

surface cauldron subsidence
[A subsidence in a cauldron (q.v.) resulting in a surface depression.]

ta   taxite
Holmes (1928), p. 223: "A general term for volcanic rocks of clastic appearance owing to the consolidation and aggregation of more than one kind of product from the same flow. When the different consolidation products are disposed in alternating bands the resulting rock is described as eutaxite, and the structure as eutaxitic. When the aggregation resembles a breccia, the rock is described as an ataxite, and the structure as ataxitic (Loewingson-Lessing, 1891, p. 104."

te   tear-shaped bomb
"A rotational volcanic bomb shaped like a teardrop and having an 'ear' at its constricted end; from 1 mm to more than 1 cm in length." Cf.: Pele's tears.

tephra
Stamp (1961), p. 447 quotes Thorarinsson (1944), p. 447, 1956, p. 20: "The present writer has proposed that the term tephra be used as a collective term for all material ejected through the air [or vacuum] by a volcano in the same way as lava is a collective term for all molten material flowing from a crater."
Stamp (1961), p. 447: "Thorarinsson derived this term from Aristotle and introduced it in Tefrokronologiska Studier pa Island, Geografiska *Annaler*, 1944 (with English summary). He uses tephrochronological and other compounds."

tephra cone
Longwell, et al., (1969), p. 660: "A small to moderate size cone composed of tephra."

tephra flow
Longwell, et al., (1969), p. 661: "A fluidized mass of tephra, whose particles may be red hot, that flows like a liquid. A tephra flow may create a welded tuff." [An ash flow.]

tephratic lava
Schieferdecker (1959), p. 262, term 4305: Less-preferred synonym for "agglutinate" (Thorarinsson, 1955).

tephrochronology
Walker, G. P. L., in *Nature*, from Thorarinsson (1944): "Volcanic ash chronology. A method of study of volcanic areas which depends on the fact that the ash thrown out by a volcanic explosion is dispersed over a wide area, and the resulting ash layer is sensibly of the same age over its whole extent. A succession of such ash layers represents a time sequence relative to which other events, such as the outpouring of lava flows, the retreat of glaciers, or the abandonment of human habitations, can be dated."

terminal cone
[A cone built at the seashore from reactions of the melt with seawater.]

te   tertiary fumarole
Schieferdecker (1959), p. 274, term 4473:
1) "spiracles of vapour, emitting from [localized areas] that maintained a high temperature

amidst clastic ejecta, as well as upon a lava flow after each fall of rain or of snow (Perret, 1924)."
2) "vapors formed at the contact zone of a lava flow, streaming into water."
Cf. "primary fumarole," "secondary fumarole."

th   **tholoid**
Challinor (1964), p. 246: "A dome of viscous lava squeezed up, exuded, and accumulated over a volcanic orifice." (Greek: tholos, a domed or vaulted building).
Webster 3 (1967), p. 2379: "A rounded dome-shaped mass of lava rising above the surface of a lava flow or crater floor."

*   **tholoides**
See also: "rheumatitic," "rheuklastitic," "klasmatitic."

thread-lace lava
Schieferdecker (1959), p. 249, term 4127: "Foamy lava"
See "thread-lace scoria," "reticulite."

thread-lace scoria
Macdonald, G., in Hess and Poldervaart, (1967), p. 50: Dana, (1890), vol. 1, p. 163–166: "[Volcanic] material more vesicular than cinder 'in which the walls between the vesicles have mostly been disrupted and all that remains is a network of threads marking the intersection of the cells' called "reticulite" by Wentworth and Williams, (1932, p. 41)."

throat
Schieferdecker (1959), p. 241, term 4025: "Of a volcano, the upper part of the chimney when in eruption."

thunder eggs
"A geode-like body weathered out of welded tuff or lava and commonly containing opal, agate, or chalcedony."

to   **toe**
Macdonald, G., in Hess and Poldervaart, (1967), vol. 1: "The edge of a large slow-moving pahoehoe flow commonly does not advance as a unit, but rather by the protrusion of one small tongue after another. These pahoehoe 'toes' are commonly 0.1 to 1 m across, and advance only 1–2 meters before they become immobile."
See also: "lava toe."

tongue
See: "lava tongue."

toothpaste lava
Macdonald, G. in Hess and Poldervaart, (1967), vol. 1, p. 21: "Viscous protrusions of lava squeezed out through openings in the flow crust; generally small tongues are 1–2 m long and a fraction of a meter across. Many are grooved by protrusion through an irregular orifice, and commonly they show chatter marks resulting from pulsating extrusions and friction against the sides of the opening. Successive extrusions may result in a series of 'toothpaste' squeeze-ups which sometimes are curved over at the top like breaking waves."

tr   **trass**
Webster 3 (1967), p. 2432: "A light-colored volcanic tuff resembling pozzolana in composition and occurring especially on the lower Rhine where it is ground for use in a hydraulic cement."
Synonym: "terrace", "tarras," "pozzolana."

tree cast
Halliday (1966), p. 382: "A mold of a tree surrounded by lava. Usually in pahoehoe."

tree mold
Stearns (personal communication): "A hole in a lava flow, caused by lava making a cast of a tree trunk."

trigger effect
Schieferdecker (1959), p. 268, term 4397: "A critical force needed to start off an eruption, as Perret (1924) thought a certain position of sun and moon might exercise."

tu   **tube**
See "lava tube."

tube-in-tube
Halliday (1966), p. 382: "A rudimentary lava tube formed in a secondary flow inside a preexisting lava tube."

tuff
Challinor (1964), p. 252: "Any soft porous stone, particularly calcareous tuff and volcanic tuff [consolidated fine volcanic ash]; adopted from the French *tuffe* in the 16th century. In the 18th century the Italian word tufa was introduced and was used alternatively to the older

form 'tuff'. In the middle of the 19th century geologists restricted the use of 'tuff' to volcanic rock and the use of 'tufa' to calcareous rock. Some tuffs appear to be intrusive (Hughes, and others (1960) Kilroe and McHenry, (1901). It has come to be used for interbedded pyroclastic rocks of past geological ages when it is synonymous with 'volcanic ash' often a hard and nonporous rock."

† Stokes and Varnes (1955), p. 156: "Indurated pyroclastic material, consisting wholly or predominantly of fine-grained volcanic ash or dust. Some fragments of country rock, or minor amounts of fragments larger than 4 mm., may be present. Tuffs may be classed as lithic, crystal, or vitric. . . . Tuff may or may not be deposited in water, and it may be well sorted or heterogeneous; if transported and redeposited by water the product is tuffaceous sandstone or shale."
Varieties include: "explosion tuff." "ash tuff," "dust tuff," "lapilli tuff," "pumice tuff," "scoria tuff," "pozzolana," "lithic tuff," "vitric tuff," "crystal tuff," "crystal lithic tuff," "lithic crystal tuff," "stratified dry tuff," "chaotic tuff," "subaqueous tuff," "sedimentary tuff," "welded tuff," "pisolitic tuff,"

tumescence
Schieferdecker (1959), p. 267, term 4388: "Of a [caldera or] volcano (Wentworth and Macdonald, 1953), a slight uparching caused by the uptruding magma shortly before the eruptive activity starts. The phenomenon belongs to the premonitory events."

tumulus (plural "tumuli")
Macdonald, G., in Hess and Poldervaart, (1967), vol. 1, p. 32: "A domical upheaval on surfaces of pahoehoe flows; also called 'Schollendome' and 'pressure dome'. Some tumuli are nearly circular in ground plan, but most are oval. They are common on flows confined within a crater or on flows whose forward motion was obstructed by some barrier. In Hawaii, they range from less than meter to about 5 m in height, 3 to 10 m in width, and commonly to 30 or 40 m in length. p. 33: They result from buckling of flow crust near edge of flow, with upheaval aided by hydrostatic pressure of liquid lava beneath crust. Some tumuli are hollow."
Cf. "pressure ridge."

tunnel fumarole
Schieferdecker (1959), p. 275, term 4488: "A fissure fumarole of which the opening is almost

shut off by a roof of incrustation products (Zies, 1924)."

tuff ball
[A sphere of tuff.]

tuff cone
Stokes and Varnes (1955), p. 156: "A cone made up chiefly or wholly of tuff," i.e., Diamond Head in Hawaii."
Synonym: "tephra cone."

tuffisite
Holmes (1965), text p. 269: "Intrusive tuff" formed mainly from the country rocks, as distinguished from tuffs normally deposited over the surface as volcanic ash. The term was proposed by Cloos (1941) for tuff pipes in Swabia, east of Black Forest. These are Tertiary pipes containing blocks of Jurassic strata. The blocks are detached from wall by intrusion [not fallen back into the vent] after having been blown out during intensely explosive activity. They form by the process of "fluidization."

tuff dike
[A dike of tuff.]

tuff flow
"An obsolete term for 'ash flow'."

tuffite
Fisher (1966), p. 295: "A composite clastic rock composed of greater than 50% [volcanic] pyroclastic material and less than 50% [detrital] epiclastic material."
Also: "tuffogenic rock".

tuff lava
[An extrusive rock intermediate in character between tuffs and lavas produced by a flow which] contains features indicative of both pyroclastic and true lava-flow origin."
Cook (1966), p. 161: "Belief in the existence of both ignimbrites and tufflavas as two genetically distinct rocks is now shared by 'the majority of specialists in the Soviet Union and is based on [much] factual material obtained in the study of ignimbrites and tufflavas of the Kurile Islands and Kamchatka, the Far East and Middle Asia, the Urals and Central Caucasus' (Shirinian, 1963)."

tu **tuffogenic rock**
See: "tuffite."

tu **tuffo-lava**
See: "tuff-lava."

tu   **tuff palagonite**
See: "palagonite tuff."

tu   **tuff ring**
[A ring of tuff such as Fort Rock produced by ejection of tuff from a central vent and depositing it in a ring-shaped pattern.]

tu   **tuffstone**
"Type of sandstone containing volcanic fragments of sand size."

tu   **tuffaceous**
Webster 3 (1967), p. 2461: "Of, relating to or resembling tuff."

tu   **tuffaceous sandstone**
Fisher (1966), p. 294: "Epiclastic volcanic rock [volcanic material transported by running water] in which component fragments are more than 50% of rock and diameter of fragments range from 1/16 to 1 mm."
Cf. "sand-tuff."

tu   **tuffaceous shale**
Fisher (1966), p. 294: "Epiclastic volcanic rock [volcanic material transported by running water] in which component fragments are less than 50% of rock and diameter of fragments are less than 1/16 mm."
Cf. "dust-tuff."

tu   **tuya**
Mathews (1947), p. 560: "Flat-topped, steep-sided volcanoes, here named 'tuyas', situated in northern British Columbia. They consist of nearly horizontal beds of basaltic lava resting on outward-dipping beds of fragmental volcanic rocks, formed by volcanic eruptions into several 'interglacial' lakes thawed through the Pleistocene Cordilleran ice-sheet by volcanic heat."

tw   **twin crater**
Schieferdecker (1959), p. 240, term 4015: "Two craters of more of less the same dimensions, situated very near to each other and generally separated only by a low wall (p. 241), Tangkuban Prahu in Java, for example."
Less-preferred synonym: "double crater."

tw   **twin volcano**
Schieferdecker (1959), p. 239, term 3998: "Two volcanic cones, situated near to each other, having the same base, the lower parts of the mantles being united."
Less-preferred synonym: "double cone."

ub   **ubehebe**
*   Stamp (1961), p. 465: "A crater formed by the expulsion of volcanic ash, lapilli, etc., around a volcanic vent. From the Ubehebe craters of the northern end of Death Valley, California. See Cotton, 1944, and von Engeln, 1935." [q. v. A maar crater.]

ul   **ultravulcanican**
Schieferdecker (1959), term 4339: "[Volcanic] explosions in which the ejected material is solidified and nearly cold. Luminous phenomena and real bombs are wanting. Only angular blocks accompany the lapilli and ashes. These eruptions often occur in the beginning when the crater vent is opened (Lacroix, 1930)."

un   **underground cauldron subsidence**
Schieferdecker (1959), p. 234, term 3923: "A less-preferred synonym for "subterranean cauldron subsidence" (no. q.v.): "a cauldron subsidence in which the ring faults do not penetrate to the surface (E. B. Bailey, 1909)."

up   **upwelling**
Schieferdecker (1959), p. 264, term 4349: "A rise of water which originates when volcanic gases escape with little force, driving (p. 265) up the water to a height of a few metres, as e.g., during weak explosions of Anak Krakatau, the force then being only sufficient to cause a strong wave-like activation of the water above the crater (Stehn, 1929)."

va   **vapor cloud**
[A steam cloud above a vent.]

ve   **vent**
Monkhouse (1965), p. 323: "An opening in the surface of the earth's crust, through which material is forced during a volcanic eruption." See: "volcano," "volcanic vent," "fumarole," "geyser."

ve   **vent agglomerate**
[Deposits of agglomerate surrounding a volcanic vent.]

ve   **vesicle cylinder**
Macdonald, G., in Hess in Poldervaart (1967), vol. 1, p. 37: "Well known in lavas of northwest U.S.A. A cylindrical zone in the lava in which

vesicles are notably more abundant than in the surrounding rock. The cylinders range from 1 to 10 cm or more in diameter and in length up to a meter or more although most commonly they are shorter." p. 39: "Probably formed by generation of steam from underlying wet surfaces"
See also: "pipe vesicle."

ve   vesuvian
"A type of volcanic activity which is characterized by alternating paroxysmal eruptions of magma highly charged with gases, and long periods of inactivity." [A stratovolcano evolves from this volcanic activity pattern.]
See also: "Plinian" and "paroxysmal eruption."

ve   Synonym: vulcanian

ve   vesuvian-type volcano
See "vesuvian."

vi   vitric
Stokes and Varnes (1955), p. 160: "An adjective designating volcanic ejecta consisting primarily of glassy material. It may be advisable to restrict the term to fragments made up of at least 75 percent by volume of glass. In addition to the above usage the term 'vitric' may be applied to any rock having the nature or quality of glass."
Cf. "vitric-crystal."

vi   vitric ash
"Ash with 'more than 75 percent glass'."

vi   vitric-crystal
Stokes and Varnes (1955), p. 160: "Ejecta containing between 50 and 75 percent of glass (Wentworth and Williams)."

vi   vitric tuff
Ross and Smith (1961), p. 8: "A name proposed by Pirsson (1915, p. 194) for 'tuffs produced by the sudden and violent explosion of a more or less viscous magma. The explosion of gas and rupturing begins in a liquid medium; the resulting product falls as a rigid glass. In general, ash-flow tuffs are dominantly vitric or derived from originally vitric material.' As Pirsson states, such material would fall from the air as a rigid glass (ash-fall tuffs). However, where flowage occurs, the glass may retain plasticity until after deposition."

vi   vitroclastic
Cook (1966), p. 160: "Said of a structure, characteristic of fragmental glassy rocks, in which most of the particles have crescentic or vesiculate outline."

vo   volcan
Webster 3 (1967), p. 2562: "(Sp. Volcan, from Latin Vulcanus, Roman god of fire): "Volcano"
Makiyama (1954), p. 146: "Component of earth's crust made up of volcanoes and various hypabyssal igneous bodies."

vo   volcanello
"An Italian term meaning 'small volcano'. It has been applied (1) to small, active volcanic cones, usually contained within the central crater of a volcano, as Mount Nuevo of Vesuvius and the small cones in the summit crater of Etna, and (2) to low, steep-sided hills or mounds of spatter built around a central vent or along a fissure." (after Calkins, A.G.I.)
Synonym: "spatter cone."

vo   volcanic action
Challinor (1964), p. 259: "Includes anything to do with igneous action at, or immediately below . . . the earth's surface, whether or not an actual volcano is formed."

vo   volcanic agglomerate
Stamp (1961), p. 11, quotes Geikie (1920), p. 53: "Volcanic agglomerate is the name given to a coarse admixture of large and small blocks and stones set in a matrix of comminuted rock debris and grit, which may be either abundant or meagre."

volcanic ash
Stokes and Judson (1968), p. 521: "The unconsolidated, fine-grained material thrown out in volcanic eruptions. It consists of minute fragments of glass and other rock material, and in color and general appearance may resemble organic ashes. The term is generally restricted to deposits consisting mainly of fragments less than 4 mm in size. Very fine volcanic ash composed of particles less than 0.05 mm may be called volcanic dust. The induratedequivalent of volcanic ash is tuff."

volcanic association
Turner and Verhoogen (1960): "A volcanic association may, and frequently does, include intrusive rocks, but these are generally related

to a cycle of volcanic activity and have been derived from the same parent magmas as the associated strictly volcanic rocks."

volcanic avalanche
Ross and Smith (1961), p. 3: "The sudden avalanching of large amounts of any volcanic material down the slopes of a volcano. The term has been used by many authors. p. 4: Some volcanic avalanches have been described as glowing avalanches where the material in them is red hot and contains gas. Other avalanches, that occur where large quantities of water are involved, are known as lahars."

volcanic basin
[A large depression formed by volcanic or volcano-tectonic activity.]

volcanic belt
Schieferdecker (1959), p. 223, term 3770: "A distribution of volcanoes in definite zones or belts, generally on the margin of continents or within oceanic areas."
Less-preferred synonym: "volcano zone," "volcano girdle." (no q.v.)

volcanic block
"A [solid angular] fragment, more than 32 mm in diameter and consisting of lava that was ejected from a volcanic vent when it was entirely solid."

volcanic blowpiping
See "blowpipe flame."

volcanic bomb
Hatch and Rastall (1965), p. 321: "A mass of lava [over 32 mm in diameter] ejected in a fluid condition, which acquired a vesicular structure and a highly characteristic spindle shape while traveling through the air."

volcanic butte
"An isolated hill or mountain resulting from the differential weathering or erosion and consequent exposure of a volcanic neck or of a narrow, vertical igneous intrusion into overlying weaker rock; e.g., Ship Rock, New Mexico."
Cf.: "mesa-butte."

volcanic center
An area on the earth's surface where eruptive action takes place.
See: "volcanic focus," "eruption point" (no q.v.), and "eruption center" (no q.v.).

volcanic chain
Stokes and Varnes (1955), p. 160: "A group of aligned volcanoes, the arrangement apparently controlled by a single fracture [zone] (M. Billings)."

volcanic chamber
Schieferdecker, (1959), p. 228, term 3829: "Magma chamber."

volcanic cinder
[A vesiculate volcanic fragment 4 to 32 mm in diameter.]
See "cinder."

volcanic clinker
See "clinker."

volcanic cloud
Webster 3 (1967), p. 2562: "A convoluted rolling mass of partly condensed water vapor and dust that is generally highly charged with electricity and that overhangs a volcano during eruption."

volcanic cluster
Schieferdecker (1959), p. 224, term 3776: "A group of central volcanic vents formed in a restricted area (R. A. Daly), as a result of the rise of fluxing gases from many points in the more or less irregular roof of a laccolith and their exploding through the [overlying rock]. An example known from Swabia has more than 100 central vents."
Less-preferred synonym: "cluster of volcanoes," "cluster of cones" (no q.v.).

volcanic column
Schieferdecker (1959), p. 234, term 3928: "neck."
See: "lava column."

volcanic conduit
See: "volcanic vent."

volcanic cone
Stokes and Varnes (1955), p. 160: "A conical, often symmetrical hill of volcanic material, either entirely ash or cinders, entirely lava, or partly lava and partly pyroclastic debris. In any case it is apt to be intersected by dikes of lava (Lahee, 1941)."

volcanic conglomerate
Stokes and Varnes (1955), p. 160: "Sedimentary, coarse pyroclastic material containing an abundance of large, chiefly rounded, water-

worn fragments. In most cases, they result from the erosion and redeposition of old volcanic rocks, but they may also be formed by volcanic mud flows and by the action of running water on freshly fallen ejecta (Wentworth and Williams)" Fisher (1961), p. 294: "The component fragments constitute greater than 50% of rock, and diameter of fragments is greater than 1 mm."
Cf. "volcanic breccia."

volcanic crater
See: "crater."

volcanic cycle
Longwell, et al., (1969), p. 647: "Cycle of volcanic eruption repetition of similar activities during an eruption of a volcano."

volcanic debris
[Poorly sorted fragmental material produced by volcanic processes.]

volcanic dome
Williams (1932): "A steep-sided, rounded extrusion of highly viscous lava squeezed out from a volcano, and forming a dome-shaped or bulbous mass of congealed lava above and around the volcanic vent. Portions of older lavas may be elevated by the pressure of the new lava rising from below. The structure generally develops inside a volcanic crater or on the flank of a large volcano, and is usually much fissured and brecciated."
Cf. lava dome.
Synonyms: "tholoid," "dome volcano," "cumulo-dome," "cumulo-volcano."

volcanic dust
Stokes and Judson (1968), p. 521: "Pyroclastic detritus consisting of particles of dust size (<0.05 mm in diameter)" McIntosh, (1963), p. 274: "Volcanic dust is known to have spread in the stratosphere as a veil covering more than half the surface area of the globe in some instances and to have persisted in observable quantities for up to three years. The latitude zones which are sooner or later affected probably depend greatly on the latitude of injection. Such dust veils are associated with certain atmospheric optical effects (For example, Bishop's Ring). It is also probable that significant effects on atmospheric circulation and world weather are caused by the scattering of solar radiation by widespread and persistent veils."

volcanic edifice
Longwell et al., (1969), p. 661: "A feature built [up around by a volcanic vent]."

volcanic emanations
See: "emanations, volcanic."

volcanic energy
[The energy available to produce volcanic phenomena. Some major volcanic bursts cans accrue 10^{23} ergs.]

volcanic eruption
Stokes and Judson (1968), pp. 521: "The explosive or quiet emission of lava, pyroclastics, or volcanic gases at the earth's surface, usually from a volcano but [sometimes] from fissures."

volcanic fissure trough
[A graben produced by subsidence along two relatively closely spaced nearly parallel volcanic fractures.]

volcanic flow drain
Bateman (1951), p. 99: "[The feature] forms in lava flows when the outside of the lava has solidified and liquid lava in the center drains out, leaving a pipe or tunnel."
See: "lava tube."

volcanic focus
Webster 3 (1967), p. 2562: "The subterranean center of volcanic action."

volcanic gases
Krauskopf (1967), p. 458: "In Kilauea, water is the chief gas, CO_2 is next, SO_2 third. Small amounts of CO, H_2, S_2, SO_3, Cl_2 are also present; . . . H_2S may occur in place of SO_3." p. 459: "At Katmai in Southwest Alaska, H_2O is about 99%; the others are: HCl, CO_2, HF, H_2S, N_2, with minor CO and CH_4 but no SO_2.", p. 461: In general: "water vapor is the most dominant gas, often amounting to 90% or more of the total." Next in order of abundance are carbon gases (CO_2, CO, COS, CH_4), with CO_2 dominant; next are sulfur gases (SO_2, H_2S, SO_3), with sulfur dioxide usually important at high temperatures, H_2S at low temperatures; HCl is the commonest of halogen gases."
Krauskopf (1967), p. 462: "HF is usually not determined" (where reported it is generally less than one-tenth that of HCl); "nitrogen is present chiefly as the free element, but a little NH_3 is sometimes reported; NH_4Cl is a com-

mon sublimate; "free hydrogen is usually not abundant."

volcanic glass
Nelson and Nelson (1967), p. 400: "Igneous rock, mostly extrusive, in which the magma cooled and solidified very rapidly". p. 401: "The glass occurs as more or less separate masses, or as chilled edges of minor intrusions. Examples are obsidian and pitchstone."
Synonym: "glassy rock."

volcanic graben
[A downdropped block in a volcanic province.] See: "sector graben" and "volcano-tectonic depression."

volcanic gravel
Fisher (1966), p. 295: "Term used by Vlodavetz et al., (1962), for unconsolidated pyroclastic debris with size range 2 to 10 mm—between coarse ash and lapilli."
Synonym: "gravel."

volcanic island
Udintsev, G. B. in Runcorn (1967), vol. 1, p. 656: "An 'oceanic island' consisting of tops of volcanoes which rise up above the level of the water."
See also: "island volcano."

volcanicity
Monkhouse (1965), p. 324: "The processes by which solid, liquid or gaseous materials are forced into the earth's crust and/or escape onto the surface. This includes igneous activity generally, besides that popularly associated with volcanoes."
Synonym: (obsolete) "vulcanicity," "vulcanism."

volcanic lake
[Three types of 'lake basins produced by volcanic activity' include]:
1) Circular lake occupying volcanic craters and calderas—e.g., Crater Lake in Oregon.
2) Lake originating when lava flow dams a valley, or when growth of volcanic cones obstructs a preexisting drainage system.
3) Lake resulting when solidified crust of a new lava flow collapses after still-fluid lava beneath has drained away." (after H. Zumberge).

volcaniclastic
Fisher (1961), p. 1409–1414: "A term used to encompass the entire field of clastic rocks composed in part of, or entirely of, volcanic frag-

ments." Term is "used to pinpoint and include the entire spectrum of fragmental volcanic rocks formed by any mechanism or origin, emplaced in any physiographic environment (on land, under water or under ice), or mixed with any other volcaniclastic type or with any non-volcanic fragment types in any proportion."

volcanic mountain
Longwell et al., (1969), p. 661: "A conical accumulation of volcanic materials."

volcanic mud
Stokes and Varnes (1955), p. 161: "Mud formed by the mixture of water with volcanic ash or other fragmental products of eruption. This mixture may form at the time of eruption and may then be hot and flow in streams like lava; the mud may also form later at any stage of the history of the volcanic ash. It may become a mud on addition of water on dry land or by falling into a body of water."

volcanic mudflow
Macdonald, G., in Hess and Poldervaart (1967), p. 25: "May be formed by ejection of a crater lake onto the debris-covered outer slopes of a volcanic cone, by destruction of the crater wall releasing the water of a crater lake, or melting of ice and snow by (p. 26) eruption of hot material onto or under them (as in the case of the famous 'Jokullaup' of Iceland); but by far the commonest cause is simply heavy rain . . . falling on the loose material on the flanks of the mountain."
Synonym: "lahar."

volcanic neck
Stokes and Varnes (1955), p. 161: "A column of igneous rock formed by congelation of lava or the consolidation of volcanic breccia, etc., in the conduit of a volcano. The neck may later be left standing above the adjacent country by the removal of surrounding rocks by erosion (Merriam-Webster)."
See: "volcanic plug," "volcanic tuff."

volcanic nunatak
See: "steptoe."

volcanic pipe
Challinor (1964), p. 260: "Each volcanic chimney, by which vapour, ashes or lava are discharged at the surface, may be conceived to descend in a more or less nearly vertical direction until it reaches the surface of the lava

whence the eruptions proceed. After the cessation of volcanic activity, this pipe will be left filled up with the last material discharged, which will usually take the form of a rudely cylindrical column reaching from the bottom of the crater down to the lava reservoir (Geikie, 1897). By now, owing to prolonged denudation, vast amounts of the volcanic materials have disappeared and the broad pipes of former volcanoes are revealed (Richey, 1961)."

volcanic plain
Stokes and Varnes (1955), p. 161: "A broad flat surface on lava or volcanic ejectamenta."

volcanic plug
"This term is sometimes used as a synonym of 'volcanic neck' but is restricted by some to necks consisting of a monolithic mass of solidified igneous rocks."
See: "volcanic neck."

volcanic pressure
Schieferdecker (1959), p. 229, term 3854: "Magmatic pressure."

volcanic province
"A large region of volcanic activity."

volcanic quicksand
Schieferdecker (1959), p. 273, term 4456: "A deposit, mainly composed of fine ash at an elevated temperature, in consequence of which there is maintained such a dilation of the interstitial gas as to give the whole a quite extraordinary degree of potential mobility (Perret, 1924)."

volcanic rain
Schieferdecker (1959), p. 269, term 4420: "Eruption rain."

volcanic rent
Schieferdecker (1959), p. 247, term 4495: "A violent separation of the crater floor or of the flank of the volcano, caused by volcanic pressure."

volcanic rocks
Stokes and Varnes (1955), p. 161: "Rocks that have issued from vents at the earth's surface either ejected explosively (pyroclastic material or extruded as lava, together with the near-surface dikes, sills, and plugs which form a part of the volcanic structure. Such rocks, because they solidify by rapid cooling from a high temperature and with some loss of volatile constituents that crystallized before extrusion may be present as conspicuous large crystals (phenocrysts) in the finer matrix. Gas bubbles and fluidal or breccia textures are common."

volcanic rubble
[Poorly sorted volcanic ejecta ranging from dust to blocks.] (>0.05 mm to > 1 meter)

volcanics
[A colloquial term for "volcanic rocks."]

volcanic sand
Schieferdecker (1959), p. 256, term 4227: "Unconsolidated small fragments of volcanic material with a diameter from 0.05 to 2 mm, formed by explosions [thus falling within the ash size category] (>0.05 <4 mm)."

volcanic sector graben
Schieferdecker (1959), p. 245, term 4068: "A [linear] collapse area on the flank of a volcanic structure with walls [bounded by steeply dipping faults]."

volcanic sediment
"Air or water elutriated, layered volcanic ejecta or detritus."

volcanic sequence
Schieferdecker (1959), p. 254, term 4196: "A succession of volcanic rocks, the composition of which has undergone change during the active period of a volcano (Geikie, 1897)."

volcanic shield
Longwell, et al., (1969), p. 662: "A broad convex mound of extrusive igneous rock, having surface slopes of only a few degrees."

volcanic shield cluster
[A group of volcanic shield structures as on Hawaii.]

volcanic sink
Stokes and Varnes (1955), p. 161: "Roughly circular or oval depressions which may be formed in volcanic regions by broad-scale downfaulting of the land surrounding an old vent. The settling may be caused by subterranean removal of lava which had served as a support (Lahee, 1941)." [Or by explosivity. See "caldera", which differs from some sinks by being polygenetic and over a kilometer in diameter using the definition suggested in this glossary.]

volcanic spine
Monkhouse (1965), p. 289: "A solidified mass of viscous lava, forced out by extrusion and congealed."

volcanic summit graben
Schieferdecker (1959), p. 245, term 4067: "Straight-sided depression, more or less rectangular or triangular, on the summit of a volcano [or caldera] (Bemmelen, 1949, H. Williams, 1941); obviously there are many gradations between these and calderas."

volcanic thunderstorm
Schieferdecker (1959), p. 269, term 4423: "Violent lightning and thunder phenomena accompanying the electric discharges [in steam clouds formed by volcanic eruptions] which are possibly caused by electrostatic friction of the ash in the vent."

volcanic vent
Stokes and Varnes (1955), p. 161: "The opening of a volcanic structure through which lava or ash reaches the surface."
Also: "cylindrical vent," "fissure vent," "principal vent," and "adventive vent."
Synonym: (no q.v.), "volcanic chimmey," "volcanic orifice," "volcanic channel," "channel of ascent," (q.v.) "conduit," "pipe," "diatreme," "feeding vent," "eruption channel."

volcanic wind
Webster 3 (1967), p. 2562: "A wind associated with a volcanic outburst and due to the eruption or to convection currents over hot lava."

volcanic wreck
Schieferdecker (1959), p. 247, term 4106: "Sapper's (1927), term for the 'last stage of destruction by erosion, the cone or dome shape being destroyed.' An arrangement of radial ridges or a centrifugal pattern of consequent valleys may survive (Cotton, 1944)."
Synonym: "volcanic skeleton," (no q.v.), Cf. "furrowed volcano." "cleaved volcano."

volcanism
[The skin effect of defluidization (q.v.)] Longwell, et al., (1969) p. 662: "A term designating the aggregate of processes associated with the surface phenomena involved in the transfer of materials from the Earth's interior to its surface"

*   volcanist
"Obsolete equivalent for 'volcanologist'; one

versed in the study of volcanic phenomena."

volcanite
Rittmann in Runcorn (1967), vol. 2, p. 822: "A volcanic rock."

*   volcanity
Schieferdecker (1959), p. 220, term 3722: Synonym for volcanicity.

volcanize
Webster 3 (1967), p. 2562: "To subject to or cause to undergo and be affected by volcanic heat."

volcano
From minor Roman deity, Vulcan, the god of fire and patron of blacksmiths, whose forge was beneath Mount Etna (Cottingham, 1951, p. 156–157). Plural volcanoes.
Wyckoff: "Any vent through which volcanic materials, lava, ash, nuée ardentes, etc., "are erupted; or the hill or mountain formed by the accumulation of erupted materials around the vent" "Volcanoes are sometimes classified, in fact, according to the type of eruption to which they are subject: e.g., (in order of increasing explosiveness) Hawaiian Strombolian (after Stromboli), Vulcanian (after Vulcano), Vesuvian, Peléan"

volcano mantle
Schieferdecker (1959), p. 242, term 4037: "The part of the volcano rising above the surface of the land and situated between the vent and the flank of the cone."

volcano-tectonic depression
Bemmelen (1949): "Great tectonically controlled subsidence structures caused by removal of magma from the chamber by external volcanism, followed by caving of overlying crustal strata or volcanic structures by both volcanic activity and by tectonic processes; a transition between a caldera and graben. Examples are the Ungaran volcano near Semarang, central Java; and the Toba Trough in North Sumatra."

volcano-tectonic collapse
See: "volcano-tectonic depression."

volcano-tectonic horst
Schieferdecker (1959), p. 244, term 4063: "A horst-like roof of a laccolithic intrusion, e.g., Ischia according to A. Rittmann."

volcano-tectonic subsidence structure
Schieferdecker (1959), p. 244, term 4064: "A subsidence structure partly to be explained by volcanic activity and partly by tectonic processes." See also "volcano-tectonic trough," "volcanic summit graben," "volcanic sector graben," "volcano-tectonic depression," or "caldera."

volcano-tectonic process
"[Major processes involving the upper mantle] able to produce vertical movements of the crust."

volcano-tectonic trough
Hutchinson, (1957), p. 35: "A volcano-tectonic basin for which it is difficult to distinguish between effects of volcanism and of other types of earth movement."

volcanogenic
[Originating by volcanic processes.]

volcanology
"The branch of geology that treats of volcanic phenomena." Adj. "volcanological," "vulcanological" (obsolete).

volcanologist
"One who studies occurrence, origin, and activity of volcanic structures, origin of volcanic rocks, and ore-forming volcanic processes occurring in the earth."

vu   vulcan
*   Webster 3 (1967), p. 2566: "Obsolete term for volcano."

vulcanian
"A type of eruption from a volcano whose lava is rather viscous and pasty, which quickly crusts over after an eruption. Gases accumulate beneath the congealed cover and blow off at longer and longer intervals, with correspondingly greater violence. The resulting volcanic clouds are dark, and characteristically assume a convoluted or 'cauliflower' shape as they ascend or expand. Major eruptions of many volcanoes start with a vulcanian phase where an obstructed vent causes exceptionally high pressures to be developed before the eruption can proceed normally."

*   vulcanist (obsolete)
"Volcanologist (q.v.)."

*   vulcanicity
Stamp (1961), 1.477 quotes Monkhouse (1954), p. 43: "The term vulcanicity includes in its widest sense all the processes by which solid, liquid or gaseous materials are forced into the earth's crust or escape on to the surface." Synonym: "volcanicity."

*   vulcanism
Stamp (1961), p. 477: "The Lipari Islands, in the Mediterranean between Sicily and Italy were formerly known as the Vulcanian Islands and the word 'vulcanian' is sometimes used of a type of volcanic eruption observed in the volcanoes there." See: "volcanism."

vulcanite
Extrusive rocks (from Johannsen, 1931, vol. 1, 4): Term proposed by: Scheerer: 1862, p. 138.
[Modern term is volcanic rocks or acceptably volcanics.]

*   vulcano
Webster 3 (1967), p. 2566: "Obsolete term for volcano."

*   vulcanology
See: "volcanology."

*   vulcanorium
Gakkel and Dibner, in Runcorn (1967, p. 161): "The Arctic vulcanorium, is the northernmost part of the mid-oceanic ridge in the Arctic Ocean of broad area with an axial zone and troughs along its flanks" [Axis of volcanic activity; preferred spelling "volcanorium," if term accepted.]

*   vulcano type
"In volcanology, the activity characterized by production of ash clouds (Daly, 1938)."
See also: "vulcanian."

water cupola
Schieferdecker (1959), p. 265, term 4350: "A cup-shaped vault, formed on the surface of the sea above a submarine explosion, just before the gases, bombs and ashes break through. In the beginning the expanding gases lift up the water in a dome with rounded top; later on, a truncated cone shape or even a cylinder is formed. In 1928, Stehn (1929), observed water cupolas of a height of more than 26 m and a width of 100–200 m at the base."
Less-preferred synonym: "water dome"

water dome
See: "water cupola."

water fountain
Schieferdecker (1959), p. 265, term 4353: "A spring of water:
1) The initial effect of an explosion, visible above the surface of the water, in which case no great height is reached. A better term would be a 'spurt of water' (no q.v.) (Stehn, 1929).
2) The natural sequel from upwelling water or a water cupola to the escape of gases without any ejection of solid material. The water column rises high with loss in breadth. Mud generally appears at the base of the fountain."

welded pumice
Ross and Smith (1961), p. 8: "Collapsed pumice has been called welded pumice by Iddings (1899)."

welded texture
Bayly (1968), p. 41: "Rocks that look as if they are a mass of separate pieces that have been piled together while still plastic may be called welded, especially if the constituents are largely glassy. Such rocks are supposed to form by deposition of ejecta while they are still hot enough to be deformable and to stick to one another. Gas-charged ash flows are thought to be the main generators of welded rocks."

welded tuff
Ross and Smith (1961), p. 8: "Mansfield and Ross (1935, p. 308, 321) described welded volcanic tuffs, that is, those in which individual fragments had remained plastic enough to become partly or wholly welded. In a few specimens the original forms are unmodified; in others there is flattening, but without obliteration of characteristic ash structure; and in a few, extreme flattening and slight flowage has almost obscured the original structure. The term 'welded tuff' is self explanatory, and has been widely used by many geologists." [Welded tuffs occur in consolidated ash or pyroclastic flows.] See: "ignimbrite."

welding
Ross and Smith (1961), p. 8: "Iddings is widely recognized as discovering the phenomenon of rock fragment fusing or welding . . . although the term 'welded' was used long before by Zirkel (1876, p. 267)."

wi   wilsonite
Holmes (1928), p. 240: "A rhyo-andesite tuff containing fragments of pumice and andesite in a matrix consisting of shreds of glass in a granular isotropic base. The rock has also been interpreted as a brecciated rhyolite flow, but the evidence appears to be against this view."

z   zone, volcanic
See: "volcanic belt."

REFERENCES FOR
THE GLOSSARY

Allen, E. J. and Zies, E. G. 1923. A chemical study of the fumaroles of the Katmai region. Contrib. Tech. Papers, Katmai Ser., no. 2, p. 75–155.

Aramaki, S., and Yamasaki, S. 1963. Pyroclastic flows in Japan. *Bull. Volcanol.* 26: 89–99.

Bailey, E. B. 1909. The cauldron subsidence of Glen Coe and the associated igneous phenomena. Geol. Soc. (London) *Quart. J.* 65: 611–678.

Bailey, E. B. 1926. Domes in Scotland and South Africa. *Geol. Mag.* vol. lxiii, pp. 481–495.

Bailey, D. L. 1929. *An etymological dictionary of chemistry and mineralogy.* London: Arnold, 307p.

Bateman, A. M. 1951. *Economic mineral deposits,* 2nd ed. New York: John Wiley, 916pp.

Bayly, Brian. 1968. *Introduction to petrology.* Englewood Cliffs, New Jersey: Prentice-Hall, 371pp.

Bemmelen, R. W. van. 1949. *The Geology of Indonesia,* vol. 1. The Hague: Government Printing Office, pp. 191–203.

Bemmelen, R. W. van, and Rutten, M. G. 1955. *Tablemountains of northern Iceland.* Leiden: E. J. Brill.

Blyth, F. G. H. 1940. The nomenclature of pyroclastic deposits. *Bull. Volcanol.* ser. 2, 6: 145–156.

Berry, J. A. 1929. *Mineralogical Mag.* vol. 22.

Broderick, T. M. 1936. Differentiation in lavas of the Michigan Keweenawan. *Geol. Soc. Am. Bull.* 46: 503–508.

Brown, V. J., and Runner, D. G. 1939. *Engineering terminology* 2nd ed. Chicago: Gillette, 439pp.

Bullard, F. M. 1953. Condition of active volcanoes of Italy in 1952. *Volcano Letter* (Hawaii Univ.) 521: 1–5.

Carlisle, D. 1963. Pillow breccias and their aquagene tuffs, Quadra Island, British Columbia. *J. Geol.* 71: 48–71.

Challinor, J. 1964. *A dictionary of geology.* 2nd ed. New York: Oxford University Press, 289pp.

Cloos, 1941. Bau und Tätigkeit von Tuffschloten; Untersuchungen an dem Schwäbischen Vulkan. *Geol. Rundschau* Bd. 32, H. 6–8, pp. 709–800.

Coats, R. C. 1968. The Circle Creek Rhyolite, A volcanic complex in Northern Elko County, Nevada. *Geol. Soc. Am. Mem.* 116: 69–106.

Cook, E. F. 1966. Paleovolcanology: *Earth-Sci. Rev.* 1: 155–174.

Cottingham, K. 1951. The geologist's vocabulary. *Sci. Monthly* 72: 154–163.

Cotton, C. A. 1944. *Volcanoes as landscape forms.* Christchurch, New Zealand: Whitcombe and Tombs, 416pp.

Curtis, G. H. 1954. Mode of origin of pyroclastic debris in the Mehrten formation of the Sierra Nevada. *Calif. Univ. Pub. Dept. Geol. Sci. Bull.* no. 9, 29: 453–502.

Daly, R. A. 1914. *Igneous rocks and their origin.* New York: McGraw-Hill Book Co., 563pp.

Daly, R. A. 1938. *Architecture of the earth*: New York: Appleton Century Co., 211pp.

Daly, R. A. 1933. *Igneous Rocks and the depths of the Earth.* New York: McGraw-Hill Book Co., 598pp.

Dana, J. D. 1890. *Characteristics of volcanoes.* New York: Dodd Mead and Co. 399pp.

Davis W. M. 1930 The Peacock Range Arizona. *Geol. Soc. Am. Bull.* vol. 41.

Dell'Erba, 1892. Considerazioni sulla genesi del piperno: Gior. Mineralogia, *Cristallografia and Petrografia* 3: 23–53.

Escher, B. G. 1920. L'eruption du Galounggoung en Juillet 1918. *Natuurk. Tijdschr. Ned. Ind.* (Batavia) 80: 260–264.

Escher, B. G. 1929. On the formation of caldera. *Leisdche Geol. Mededeelingen, Deel III*, Aflevering 4, Leyden, pp. 184–219.

Fay, A. H. 1920. A glossary of the mining and mineral industry. *U. S. Bur. Mines, Bull.* 95, 754pp.

Fenner, C. N. 1923. The origin and mode of emplacement of the great tuff deposit in the Valley of Ten Thousand Smokes. *Natl. Geograph. Soc.*, Contrib. Tech. Papers, Katmai Ser., no. 1, 74pp.

Fenner, C. N. 1937. Tuffs and other volcanic deposits of Katmai and Yellowstone Park. *Am. Geophys. Union Trans.*, 18th Ann. Mtg. pt. 1, pp. 236–239.

Fenner, C. N. 1948. Incandescent tuff flows in southern Peru: *Geol. Soc. Am. Bull.* 59: 879–893.

Finch, R. H. 1933. Block lava. *J. Geol.* 41: 769–770.

Fisher, R. V. 1966. Rocks composed of volcanic fragments and their classification. *Earth-Sci. Rev.* 1: 287–298.

Fisher, R. V. 1961. Proposed classification of volcaniclastic sediments and rocks. *Geol. Soc. Am. Bull.* vol. 72, 1409–1414.

Fritsch, K. V., and Reiss, W. 1868. *Geologische beschreibung der Insel Tenerife.* Winterthur: Verlag von Wurster & Co., 494pp.

Geikie, A. 1897. *Ancient volcanoes of Great Britain.* 2 vols. London: The MacMillan Co. 955pp.

Geikie, A. 1902. *Textbook of geology.* London: The MacMillan Co.

Geikie, A. 1920. Structural and field geology, 4th ed. Edinburgh: Oliver & Boyd.

Geze, B. 1943. La Montagne Noire. *France Serv. Carte Géol.* B. (No. 212) 44: 253–257.

Gregg, D. R. 1956. Eruption of Ngauruhoe 1954–1955: *New Zealand J. Sci. Tech.* Sec. B. 37: 675–688.

Griggs, R. F. 1922. *The Valley of Ten Thousand Smokes [Alaska]* Washington: Nat. Geograph. Soc. 341pp.

Halliday, W. R. 1966. *Depths of the earth; caves and caverns of the United States.* New York: Harper & Row, 398pp.

Hatch, F. H., and Rastall, R. H. 1965. *Petrology of the sedimentary rocks*, 4th ed. revised by J. Trevor Greensmith. London: Thomas Murby, 408pp. (*Texbook of petrology*, vol. 2).

Haug, E. 1907. *Traité de Geologie: vol. 1 Les phenomènes geologique.* Paris: Armand Coln., 538pp.

Hess, H. H., and Poldervaart, A., eds. 1967. *Basalts: the Poldervaart treatise on rocks of basaltic composition*, vol. 1, New York: Interscience, 495pp.

Himus, G. W. 1955. *A dictionary of geology.* Baltimore: Penguin Books.

Hobbs, W. H. 1919. *Earth features and their meaning*, New York: The MacMillan Co.

Hobbs, W. H. 1953. *Origin of the lavas of the Pacific region.* Pacific Sci. Congr., 7th New Zealand, 1949, Proc. 2: 346–357.

Holmes, A. 1928. *The nomenclature of petrology*, 2nd ed. London: Thomas Murby, 284pp.

Hughes, C. J. 1960. The Southern Mountains igneous complex, Isle of Rhum (with discussion). Geol. Soc. (London) *Quart. J.* 116: 111–138.

Hutchinson, G. E. 1957. *A treatise on limnology.* New York: John Wiley and Sons.

Iddings, J. P. 1899. *Geology of the Yellowstone National Park.* U. S. Geol. Survey Mon. 32, pt. 2, chap. 10, pp. 356–430.

Jagger, T. A. 1917. On the terms aphrolith and dermolith. *J. Washington Acad. Sci.* 7: 277–281.

Jagger, T. A. 1931. Lava stalactites, toes, and 'squeeze-ups' *Volcano Letter* 345, 1–3.

Jagger, T. A. 1947. Origin and development of craters [Hawaian volcanoes]. *Geol. Soc. Am. Mem.* 21, 508pp.

Jenks, W. F., and Goldich, S. S. 1956. Rhyolitic tuff flows in southern Peru: J. Geol. 64: 156-172.

Johannsen, A. 1939. *Introduction, textures, classification and glossary*, 2nd ed. Chicago: University of Chicago Press, 318pp. (A descriptive petrography of the igneous rocks, vol. 1).

Johnston-Lavis, H. J. 1886. On the fragmentary ejecta of volcanoes. *Proc. Geol. Assoc.* 9: 421–432.

Jones, A. E. 1943. Classification of lava surfaces. *Am. Geophys. Union Trans.* pt. 1: 265–268.

Jung. J., and Brousse, R. 1959. *Classification modale des rocks eruptives, utilisant les données fournies par le compteur de points*. Paris: Masson et Cie, 122pp.

Kemp, J. F. 1940. *A Handbook of Rocks*, 6th ed. revised by F. Grout. New York: D. Van Nostrand & Co. 300pp.

Kilroe, J. R., and McHenry, A. 1901. On intrusive tuff-like, igneous rocks and breccias in Ireland: Geol. Soc. (London) *Quart. J.* 57: 479–489.

Kotô, B. J. 1916. The great eruption of Sakurajima in 1914. Tokyo: *J. Coll. Sci. Imp. Univ.* vol. 38.

Kozu, S. 1934. The great activity of Komagatake [Japan] in 1929. *Tschermaks Mineralog. u. Petrog. Mitt.* 45: 133–174.

Krauskopf, K. B. 1967. *Introduction to geochemistry*. New York: McGraw-Hill book Co., 721pp.

Kuno, H. 1941. Characteristics of deposits formed by pumice flows and those by ejected pumice *Tokyo Univ. Earthquake Res. Inst. Bull.* pt. 1, 19: 144–149.

Lacroix, A. 1936. Composition mineralogique des rocks volcaniques de l'Ile de Pâques: *Acad. Sci.*, [Paris] *Compt. Rend.* 202: 527–530.

———, 1903. L'éruption de la montagne Pelée en Janvier 1903: *Acad. Sci.* [Paris] *Compt. Rendu.* 136: 442–445.

——— 1903. Principaux résultats de la mission de la Martinique: *Acad. Sci.* [Paris] *Compt. Rendu.* 136: 871–876.

——— 1930. *Remarques sur les matériaux de projection des volcans et sur la genèse des roches pyroclastiques qu'ils constituent*. Soc. Géol. France Livre Jubilaire Centenaire. 1830–1930. 2: 431–472.

Lahee F. H. 1941. *Field geology*. 4th ed. New York: McGraw-Hill Book Co. 853pp.

Lapparent A. de. 1906. *Traité de géologie*. Paris: Masson et Cie 1961pp.

Loewingson-Lessing, F. 1911. The fundamental problems of petrogenesis. *Geol. Mag.* 8: 248–257.

Longwell. C. R., *et al.* (1969). *Physical geology*. New York: John Wiley and Sons, 685pp.

Lyell, Sir Charles 1871. *Principles of geology*, 11th ed. New York: 671 plages.

Macdonald, G. A. 1939. An intrusive pépérite at San Pedro Hill, California *Univ. Calif. Pub. Bull. Dept. Geol. Sci.* 24: 329–338.

Macdonald, G. A. 1953. Pahoehoe, aa, and block lava. *Am. J. Sci.* 251: 169–191.

Makiyama, J. 1954. Syntectonic construction of geosynclinal neptons: *Kyoto Univ. Sci., Mem.*, s.B., no. 2, 21, 115–149.

Mansfield, G. R., and Ross, C. S. 1935. Welded rhyolitic tuffs in southeastern Idaho: *Am. Geophys. Union Trans.*, 16th Ann. Mtg., pt. 1, pp. 308–321.

Marshall, G. 1932. Notes on some volcanic rocks of the North Island of New Zealand. *New Zealand J. Sci. Technol.* B. 13(4): 198–202.

Mathews, L. 1947. Tuyas, flat-topped volcanoes in northern British Columbia: *Am. J. Sci.* 245: 560–570.

Matumoto, T. 1943. The four gigantic caldera volcanoes of Kyusyu: *Japan. J. Geol. Geography*. vol. 19, spec. no. 57pp.

McIntosh, D. H. 1963. *Meteorological glossary*. 4th ed. London: Her Majesty's Stationery Office, 288pp.

Mercalli, G. 1907. *I vulcani activi della terra.* Milan: Hoepli.

Monkhouse, F. J. 1954. *The principles of physical geography.* London: Univ. of London Press.

Monkhouse, F. J. 1965. *A Dictionary of Geography.* Chicago: Aldine, 344pp.

Nelson, A., and Nelson, K. D. 1967. *Dictionary of applied geology, mining and civil engineering.* New York: Philosophical Library, 421pp.

Neuman van Padang, M. 1932. *Die Beziehungen zwischen Anfang und grösster Kraftentfaltung eines vulkanischen Ausbruches.* Handel. zesde Ned. Ind. Nat. Congr., 1932, Geol. Sec. 26, Sept. 1931, Bandoeng, 1932, pp. 670–679.

Nichols, R. L. 1936. Flow units in basalt. J. *Geol.* 44: 617–630.

Pantó, G. 1962. The role of ignimbrites in the volcanism of Hungary. *Acta Geol. Acad. Sci. Hungary* 6: 307–331.

Pantó, G. 1963. Ignimbrites of Hungary with regard to their genetics and classification. *Bull. Volcanol.* 25: 174–181.

Perret, F. A. 1924. *The Vesuvius eruption of 1906.* Washington: Carnegie Inst., Pub. 339, 151pp.

Pirsson, L. V. 1915. The microscopical characters of volcanic tuffs—a study for students: *Am. J. Sci.* ser. 4, 40: 191–211.

Reck, H. 1915. Physiographische Studie über vulkanische Bomben. *Zeit. Vulkanologie,* 1914–1915, Erganzungsband, 124pp.

Reck, H. 1930. Die Masseneruptionen: unter besonderer Würdigung der Arealeruption in ihrer systematischen und genetischen Bedeutung für das islandische Basaltdeckengebirge: *Deutsche Islandsforschung* 1930, 2, Breslau, pp. 24–49.

Playfair, Sir J. 1802. *Illustrations of the Huttonian theory of the earth:* Edinburgh: Cadell, Davies and William Creech, 528pp.

Reynolds, D. L. 1954. Fluidization as a geological process, and its bearing on the problem of intrusive granites. *Am. J. Sci.* 252: 577–613.

Richey, J. E. 1961. *British regional geology: The Tertiary volcanic districts.* 3rd ed. H. M. Geological Survey.

Rittmann, A. 1933. Die geologisch bedingte Evolution und Differentiation des Somma—Vesuvius magmas. *Zeit. Vulkanologie,* bd. 15, h. 1–2, pp. 8–94.

Rittmann, A. 1944. *Vulcani attivita e genesi.* Napoli: Edetrice Politecnica.

Rittmann, A. 1958. Cenni sulle colate di ignimbriti: *Catania Accad. Gioenia Sci. Naturali Boll.* 4: 524–533.

Rittmann, A. 1960. *Vulkane und ihre Tätigheit.* Stuttgart: Ferdinand Enke.

Rosenbusch, H. 1887. *Mikroskopische Physiographie, II.* Stuttgart: O. Mügge.

Ross, C. S., and Smith, R. L. 1961. *Ash-flow tuffs: their origin, geologic relations, and identification.* Washington: Government Printing Office, 81pp. (U. S. Geol. Survey, Prof. Paper 366).

Runcorn, S. K., ed. 1967. *International dictionary of geophysics:* New York: Pergamon Press, Inc., vol. 1–2, 1728pp.

Russel, I. C. 1897. *Volcanoes of North America,* New York: 346pp.

Sapper, K. 1919. Über Hornitos und verwandte Gebilde: *Zeit. Vulkanologie* 5: 1–39.

Sapper, K. 1927. *Vulkankunde:* Stuttgart: J. Engelhorns Nachf., 424pp.

Scheerer, T. 1862. *Die Geneuse des sachsischen Erzebirges und verwandte Gesteine:* nach ihrer chemischen Constitution und geologischen Bedeutung. Z.d.g.G., XIV.

Schieferdecker, A.A.G., ed. 1959. *Geological nomenclature.* Gorinchem: Royal Geol. and Mining Soc. of the Netherlands, 523pp.

Schneider, K. 1911. *Die Vulkanischen Erscheinungen der Erde,* Berlin:

Sharpe, W. and Stewart, C. F. 1938. *Landslides and related phenomena:* New York: Columbia Univ. Press.

Shirinian, K. G. 1963. *Hyaloclastites and conditions of their formation in Armenia.* Tr. Lab. Paleovulkanol. Kazakhsk. Nauchn.—Issled. Inst. Mineral'n Sur'ya. 2: 200–210.

Smith, R. L. 1960. Ash flows: *Geol. Soc. Am. Bull.* 71: 795–842.

Stamp, L. D., ed. 1961. *A glossary of geographic terms.* New York: John Wiley & Sons, 539pp.

Stehn, Ch. E., 1928. De Batoer op Bali en zijn erupties in 1926. *Vulk. en Seis.* Meded.

9. Bandoeng, pp. 1–67.

Stehn, Ch. E. 1929. *The geology and volcanism of the Krakatau group*. Fourth Pacific. Sci. Congr. Batavia, pp. 1–55.

Stokes, W. L., and Judson, S. 1968. *Introduction to geology, Physical and Historical*: Philadelphia: Fortress.

Stokes, W. L., and Varnes, D. J. 1955. *Glossary of selected geologic terms, with special reference to their use in engineering*: Denver: Colo. Sci. Soc., 165pp. (Proc. of Colo. Sci. Soc., vol. 16).

Strahler, A. N. 1963. *The Earth Sciences*. New York: Harper & Row, 681pp.

Streckeisen, A. L. 1967. Classification and nomenclature of igneous rocks. *Neues Jahrbuch für Mineralogie*. Abhandlungen, bd. 107, pp. 144–240.

Stübel, A. 1903. *Martinique und St. Vincent. Ein Studie zur wissenschaftlicher Beurteilung der Ausbrüche auf den Kleinen Antillen, 1902*. Leipzig: Max. Weg, 36pp.

Suess, E., 1902. Hot springs and volcanic phenomena: *Geograph. J*. 20: 517–522.

Swayne, J. C. 1956. *A concise glossary of geographical terms*. London: George Philip & Son, 164pp.

Tanakadate, H. 1914. Sulla conca di Bolsena: Boll. R. Comm. Geol., 44: 135–155.

Taverne, N.J.M. 1926. Vulkaan-studien op Java. *Vulk. en Seism. Medeed.*, VII.'s-Gravenhage, Algemeene landsdrukkerij, 132pp.

Thorarinsson, S. 1944. Tefrokronologiska Studier på Island. *Geografiska Annaler*, (with English summary).

Thorarinsson, S. 1953a. Conference on terminology of glacier bands: *J. Glaciol*. 2: 229–230.

Thorarinsson, S. 1953b. The crater groups in Iceland: *Bull. Volcanol*. ser. 2, 14: 3–44.

Thorarinsson, S. 1955. Discussions: *Bull. Volcanol.*, ser. 2 16: 11–13.

Thorarinsson, S. 1956. *Resumenes de los Trabajos Presentados*. 20th Congreso Geologico International, Mexico, 1956.

Thornbury, W. D. (1954). *Principles of geomorphology*. New York. John Wiley & Son, 618pp.

Tomkieff, S. I. 1940. The basalt lavas of the Giant's Causeway district of northern Ireland. *Bull. Volcanol*. ser. 2, 6: 89–143.

Tsuya, H. 1930. The volcano Komagatake, Hokkaido, its geology, activity, and petrography: Tokyo Univ. Earthquake Res. Inst., Bull. 8: 238–270.

Turner, F. J., and Verhoogan, J. 1960. *Igneous and metamorphic petrology*. New York: McGraw-Hill Book Co.

Vlodavetz, V. I. 1962. *Voprosy vulkanizma 1959*. Moscow: Akad Nauk SSSR, 451pp.

Tyrrell, G. W. 1928. *The principles of petrology*. London: Methuen and Co.

von Engeln, O. D. 1932. The Ubehebe craters and explosion breccias in Death Valley, California. *J. Geol*. 40: 726–734.

von Engeln, O. D. 1942. *Geomorphology: systematic and regional*. New York: The MacMillan Co., 655pp.

von Humboldt, A. 1823. *Essai Geognostique sur le Gisement des Roches dans les Deux Hemispheres*. Paris.

von Wolff, F. 1914. *Der Vulkanismus*. vol. 1, pt. 2: Stuttgart: Ferdinand Enke, 711pp.

Walker, G. P. L., and Skelhorn, R.B. 1966. Some associations of acid and basic igneous rocks. *Earth-Sci. Rev*. 2: 93–109.

Washington, H. W. 1922. Isostasy and rock density. *Geol. Soc. Am. Bull*. 33: 378–381.

Webster's 3rd new international dictionary of the English language. 1967. unabridged. Philip B. Grove, ed., Springfield, Mass: G. & C. Merriam. 2,662pp.

Wentworth, C. K., and Macdonald, G. A. 1953. Structures and forms of basaltic rocks in Hawaii: *U. S. Geol. Surv. Bull*. vol. 994, 98pp.

Wentworth, C. K., and Williams, H. 1932. *The classification and terminology of the pyroclastic rocks*. Nat. Res. Council, Div. of Geol. and Geography, Committee on Sedimentation. Report for 1930–1932, pp. 19–53 (Nat. Res. Council, Bull. No. 89).

Weyl, R. 1966. *Geologie der Antillen Beiträge zur regionalen Geologie der Erde*. vol. 4, Gebrüder Borntraeger, 410pp.

Williams, H. 1932. The dacites of Lassen Peak and vicinity, California and their basic inclusions. *Am. J. Sci.* [5] vo. 22.

Williams, H. 1932. The history and character of volcanic domes. Univ. Calif. Publ. *Dept. Geol. Sci. Bull.* 21: 51–146.

Williams, H., 1941, Calderas and their origin. Univ. Calif. Publ. 25: *Dept. Geol. Sci. Bull.* 239–346.

Williams, H. 1942. *The geology of Crater Lake National Park, Oregon.* Washington: Carnegie Inst.

Williams, H. 1954. Problems and progress in volcanology. Geol. Soc. (London) *Quart. J.* 109: 311–332.

Wright, A. E., and Bowes, D. R. 1963. Classification of volcanic breccias; a discussion: *Geol. Soc. Am. Bull.* 74: 79–86.

Zambonini, F. 1919. Il tufo pipernoide della campania e i suoi minerali: (Italy) *R. Comitato Geologico Mem.* pt. 2, 7: 1–130.

Zeis. E. G. 1924. The fumarolic incrustations in the Valley of Ten Thousand Smokes. *Nat. Geographic Soc.*, Contrib. Tech. Papers I, Katmai Ser. No. 3.

Zirkel, F. 1876. *Microscopical petrography.* U. S. Geol. Explor. 40th Parallel Rept. (King), vol. 6, 297 pages.

PRINCIPAL CALDERAS
AND
ACTIVE VOLCANOES

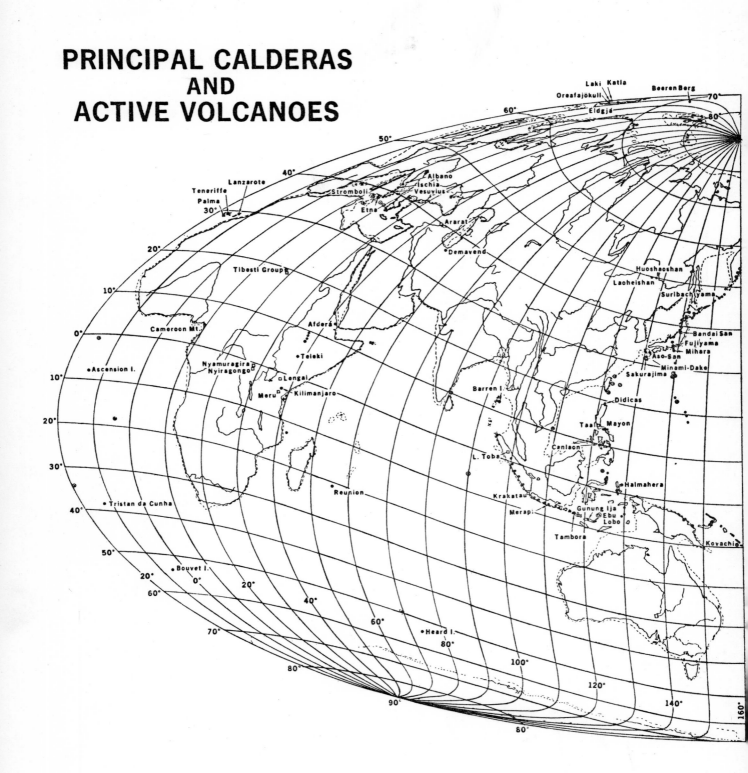

TILTED MOLLWEIDE EQUAL AREA PROJECTION
OF THE WORLD. CENTERED ON LATITUDE
20° N., AND LONGITUDE 160° E.